T0212892

Lecture Notes in Computer Science 9104

Commenced Publication in 1973
Founding and Former Series Editors:
Gerhard Goos, Juris Hartmanis, and Jan van Leeuwen

More information about this series at http://www.springer.com/series/7408

Marco Bernardo · Einar Broch Johnsen (Eds.)

Formal Methods
for Multicore Programming

15th International School on Formal Methods
for the Design of Computer, Communication,
and Software Systems, SFM 2015
Bertinoro, Italy, June 15–19, 2015
Advanced Lectures

 Springer

Editors
Marco Bernardo
Università di Urbino "Carlo Bo"
Urbino
Italy

Einar Broch Johnsen
University of Oslo
Oslo
Norway

ISSN 0302-9743   ISSN 1611-3349   (electronic)
Lecture Notes in Computer Science
ISBN 978-3-319-18940-6   ISBN 978-3-319-18941-3   (eBook)
DOI 10.1007/978-3-319-18941-3

Library of Congress Control Number: 2015938440

LNCS Sublibrary: SL2 – Programming and Software Engineering

Springer International Publishing AG Switzerland is part of Springer Science+Business Media
(www.springer.com)

Preface

This volume collects a set of papers accompanying the lectures of the 15th International School on Formal Methods for the Design of Computer, Communication and Software Systems (SFM). This series of schools addresses the use of formal methods in computer science as a prominent approach to the rigorous design of the above-mentioned systems. The main aim of the SFM series is to offer a good spectrum of current research in foundations as well as applications of formal methods, which can be of help to graduate students and young researchers who intend to approach the field. SFM 2015 was devoted to multicore programming and covered topics such as concurrency and coordination mechanisms, architecture and memory models, and type systems. The five papers of this volume represent the broad range of topics of the school.

The paper by Brandauer, Castegren, Clarke, Fernandez-Reyes, Johnsen, Pun, Tapia Tarifa, Wrigstad, and Yang presents Encore, an object-oriented parallel programming language specifically developed for supporting multicore computing and addressing performance and scalability issues. Arbab and Jongmans show how to use Reo, a language adhering to an interaction-centric model of concurrency, to coordinate multicore computing. Alglave's paper discusses concurrent programming together with the description of the execution models of the machines on which software is ran. Coppo, Dezani-Ciancaglini, Padovani, and Yoshida provide an introduction to multiparty session types, a class of behavioral types specifically targeted at describing protocols in distributed systems based on asynchronous communication. Finally, the paper by Castegren, Östlund, and Wrigstad proposes refined ownership types to reason about correctness on a local scale, for fine-grained parallelism, and for coarse-grained parallelism.

We believe that this book offers a useful view of what has been done and what is going on worldwide in the field of formal methods for multicore programming. This school was organized in collaboration with the EU FP7 project UpScale, whose support we gratefully acknowledge. We wish to thank all the speakers and all the participants for a lively and fruitful school. We also wish to thank the entire staff of the University Residential Center of Bertinoro for the organizational and administrative support.

June 2015

Marco Bernardo
Einar Broch Johnsen

Contents

Parallel Objects for Multicores: A Glimpse at the Parallel Language ENCORE

Stephan Brandauer[1], Elias Castegren[1], Dave Clarke[1(✉)],
Kiko Fernandez-Reyes[1], Einar Broch Johnsen[2], Ka I. Pun[2],
S. Lizeth Tapia Tarifa[2], Tobias Wrigstad[1], and Albert Mingkun Yang[1]

[1] Department of Information Technology, Uppsala University, Uppsala, Sweden
dave.clarke@it.uu.se
[2] Department of Informatics, University of Oslo, Oslo, Norway

Abstract. The age of multi-core computers is upon us, yet current programming languages, typically designed for single-core computers and adapted *post hoc* for multi-cores, remain tied to the constraints of a sequential mindset and are thus in many ways inadequate. New programming language designs are required that break away from this old-fashioned mindset. To address this need, we have been developing a new programming language called ENCORE, in the context of the European Project UPSCALE. The paper presents a motivation for the ENCORE language, examples of its main constructs, several larger programs, a formalisation of its core, and a discussion of some future directions our work will take. The work is ongoing and we started more or less from scratch. That means that a lot of work has to be done, but also that we need not be tied to decisions made for sequential language designs. Any design decision can be made in favour of good performance and scalability. For this reason, ENCORE offers an interesting platform for future exploration into object-oriented parallel programming.

1 Introduction

Nowadays the most feasible way for hardware manufacturers to produce processors with higher performance is by putting more parallel cores onto a single chip. This means that virtually every computer produced these days is a parallel computer. This trend is only going to continue: machines sitting on our desks are already parallel computers, and massively parallel computers will soon be readily at our disposal.

Most current programming languages were defined to be sequential-by-default and do not always address the needs of the multi-core era. Writing parallel programs in these languages is often difficult and error prone due to race conditions and the challenges of exploiting the memory hierarchy effectively. But because every computer will be a parallel computer, every programmer needs to become

---

Partly funded by the EU project FP7-612985 UPSCALE: From Inherent Concurrency to Massive Parallelism through Type-based Optimisations (http://www.upscale-project.eu).

M. Bernardo and E.B. Johnsen (Eds.): SFM 2015, LNCS 9104, pp. 1–56, 2015.
DOI: 10.1007/978-3-319-18941-3_1

a parallel programmer supported by general-purpose parallel programming languages. A major challenge in achieving this is supporting scalability, that is, allowing execution times to remain stable as both the size of the data and available parallel cores increases, without obfuscating the code with arbitrarily complex synchronisation or memory layout directives.

To address this need, we have been developing the parallel programming language ENCORE in the context of the European Project UPSCALE. The project has one ambitious goal: to develop a general purpose parallel programming language (in the object-oriented vein) that supports scalable performance. Because message-based concurrency is inherently more scalable, UPSCALE takes actor-based concurrency, asynchronous communication, and guaranteed race freedom as the starting points in the development of ENCORE.

ENCORE is based on (at least) four key ingredients: *active object parallelism* for coarse-grained parallelism, *unshared local heaps* to avoid race conditions and promote locality, *capabilities for concurrency control* to enable safe sharing, and *parallel combinators* for expressing high-level coordination of active objects and low-level data parallelism. The model of active object parallelism is based on that of languages such as Creol [21] and ABS [20]. It requires sequentialised execution inside each active object, but parallel execution of different active objects in the system. The core of the local heaps model is a careful treatment of references to passive objects so that they remain within an active object boundary. This is based on Joëlle [13] and involves so-called sheep cloning [11,29] to copy arguments passed to methods of other active objects. Sheep cloning is a variant of deep cloning that does not clone references to futures and active objects. Capabilities allow these restrictions to be lifted in various ways to help unhinge internal parallelism while still guaranteeing race free execution. This is done using type-based machinery to ensure safe sharing, namely that no unsynchronised mutable object is shared between two different active objects. Finally, ENCORE includes parallel combinators, which are higher-order coordination primitives, derived from Orc [22] and Haskell [30], that sit both on top of objects providing high-level coordination and within objects providing low-level data parallelism.

This work describes the ENCORE language in a tutorial fashion, covering course-grained parallel computations expressible using active objects, and fine-grained computations expressible using higher-order functions and parallel combinators. We describe how these integrate together in a safe fashion using capabilities and present a formalism for a core fragment of ENCORE.

Currently, the work on ENCORE is ongoing and our compiler already achieves good performance on some benchmarks. Development started more or less from scratch, which means not only that we have to build a lot of infrastructure, but also that we are free to experiment with different implementation possibilities and choose the best one. We can modify anything in the software stack, such as the memory allocation strategy, and information collected about the program in higher levels can readily be carried from to lower levels—contrast this with languages compiled to the Java VM: source level information is effectively lost in translation and VMs typically do not offer much support in controlling memory layout, etc.

ENCORE has only been under development for about a year and a half, consequently, anything in the language design and its implementation can change. This tutorial therefore can only give a snapshot of what ENCORE aspires to be.

Structure. Section 2 covers the active object programming model. Section 3 gently introduces ENCORE. Section 4 discusses active and passive classes in ENCORE. Section 5 details the different kinds of methods available. Section 6 describes futures, one of the key constructs for coordinating active objects. Section 7 enumerates many of the commonplace features of ENCORE. Section 8 presents a stream abstraction. Section 9 proposes parallel combinators as a way of expressing bulk parallel operations. Section 10 advances a particular capability system as a way of avoiding data races. Section 11 illustrates ENCORE in use via examples. Section 12 formalises a core of ENCORE. Section 13 explores some related work. Finally, Sect. 14 concludes.

2   Background: Active Object-Based Parallelism

ENCORE is an active object-based parallel programming language. Active objects (Fig. 1), and their close relation, actors, are similar to regular object-oriented objects in that they have a collection of encapsulated fields and methods that operate on those fields, but the concurrency model is quite different from what is found in, for example, Java [16]. Instead of threads trampling over all objects, hampered only by the occasional lock, the active-object model associates a thread of control with each active object, and this thread is the only one able to access the active object's fields. Active objects communicate with each other by sending messages (essentially method calls). The messages are placed in a queue associated with the target of the message. The target active object processes the messages in the queue one at a time. Thus at most one method invocation is active at a time within an active object.

Method calls between active objects are asynchronous. This means that when an active object calls a method on another active object, the method call returns immediately—though the method does not run immediately. The result of the method call is a future, which is a holder for the eventual result of the method

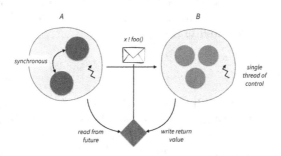

Fig. 1. Active Object-based Parallelism

call. The caller can do other work immediately, and when it needs the result of the method call, it can *get* the value from the future. If the value is not yet available, the caller blocks.

Futures can be passed around, blocked on (in various ways), or have additional functionality chained on them. This last feature, available in Javascript for instance, allows the programmer to chain multiple asynchronous computations together in a way that makes the program easy to understand by avoiding callbacks.

Actors are a similar model to active objects (though often the terminology for describing them differs). Two features are more commonly associated with active objects. Firstly, active objects are constructed from (active) classes, which typically are composed using inheritance and other well-known object-oriented techniques. This arguably makes active objects easier to program with as they are closer to what many programmers are used to. Secondly, message sends (method calls) in active objects generally return a result, via a future, whereas message sends in actors are one-way and results are obtained via a callback. Futures are thus key to making asynchronous calls appear synchronous and avoid the inversion of control associated with callbacks.

Weakness of Active Objects. Although active objects have been selected as the core means for expressing parallel computation in ENCORE, the model is not without limitations. Indeed, much of our research will focus on ways of overcoming these.

Although futures alleviate the problem of inversion of control described above in problem, they are not without code. Waiting on a future that had not been fulfilled can be expensive as it involves blocking the active object's thread of control, which may then prevent other calls depending in the active object to block. Indeed, the current implementation of blocking on a future in ENCORE is costly.

A second weakness of active objects is that, at least in the original model, it is impossible to execute multiple concurrent method invocations within an active object, even if these method invocations would not interfere. Some solutions to this problem have been proposed [19] allowing a handful of method invocations to run in parallel, but these approaches do not unleash vast amounts of parallelism and they lack any means for structuring and composing the non-interfering method invocations. For scalability, something more is required. Our first ideas in this direction are presented in Sect. 9.

3   Hello ENCORE

ENCORE programs are stored in files with the suffix .enc by convention and are compiled to C. The generated C code is compiled and linked with the ENCORE run-time system, which is also written in C. The compiler itself is written in Haskell, and the generated C is quite readable, which significantly helps with debugging.

To get a feeling for how ENCORE programs look, consider the following simple program (in file hello.enc) that prints "Hello, World!" to standard output.

```
1  #! /usr/bin/env encorec -run
2  class Main
3    def main() : void {
4      print("Hello, World!")
5    }
```

The code is quite similar to modern object-oriented programming languages such as Scala, Python or Ruby. It is statically typed, though many type annotations can be omitted, and, in many cases, the curly braces { and } around classes, method bodies, etc. can also be omitted.

Ignoring the first line for now, this file defines an active class Main that has a single method main that specifies its return type as void. The body of the method calls print on the string "Hello, World!", and the behaviour is as expected.

Every legal ENCORE program must have an active class called Main, with a method called main—this is the entry point to an ENCORE program. The runtime allocates one object of class Main and begins execution in its Main method.

The first line of hello.enc is optional and allows the compiler to automatically compile and run the program (on Unix systems such as Mac OS X). The file hello.enc has to be runnable, which is done by executing chmod u+x hello.enc in the shell. After making the program executable, entering ./hello.enc in the shell compiles and executes the generated binary, as follows:

```
$ ./hello.enc
Hello, World!
```

An alternative to the #! /usr/bin/env encorec -run line is to call the compiler directly, and then run the executable:

```
$ encorec hello.enc
$ ./hello
Hello, World!
```

4   Classes

ENCORE offers both active and passive classes. Instances of active classes, that is, *active objects*, have their own thread of control and message queue (cf. Sect. 2). Making all objects active would surely consume too many system resources and make programming difficult, so passive objects are also included in ENCORE. *Passive objects*, instances of passive classes, do not have a thread of control. Passive classes are thus analogous to (unsynchronised) classes in mainstream object-oriented languages like Java or Scala. Classes are active by default: **class** A. The keyword **passive** added to a class declaration makes the class passive: **passive class** P. Valid class names must start with an uppercase letter. (Type parameters start with a lower case letter.) Classes in ENCORE have fields and methods; there is a planned extension to include traits and interfaces integrating with capabilities (cf. Sect. 10).

A method call on an active object will result in a message being placed in the active object's message queue and the method invocation possibly runs in parallel with the callee. The method call immediately results in a future, which will hold the eventual result of the invocation (cf. Sect. 6). A method call on a passive object will be executed synchronously by the calling thread of control.

4.1 Object Construction and Constructors

Objects are created from classes using **new**, the class name and an optional parameter list: **new** Foo(42). The parameter list is required if the class has an init method, which is used as the constructor. This constructor method cannot be called on its own in other situations.

4.2 Active Classes

The following example illustrates active classes. It consists of a class Buffer that wraps a Queue data structure constructed using passive objects (omitted). The active object provides concurrency control to protect the invariants of the underlying queue, enabling the data structure to be shared. (In this particular implementation, taking an element from the Buffer is implemented using suspend semantics, which is introduced in Sect. 7.)

```
1  passive class Data { ... }
2  class Buffer
3    queue : Queue;
4
5    def init()
6      this.queue = new Queue()
7
8    def put(item : Data) : void
9      this.queue.enqueue(item)
10
11   def take() : Data {
12     while this.queue.empty() {
13       suspend;
14     };
15     this.queue.dequeue()
16   }
```

Fields of an active object are private; they can only be accessed via **this**, so the field queue of Buffer is inaccessible to an object holding a reference to a Buffer object.

4.3 Passive Classes

Passive classes in ENCORE correspond to regular (unsynchronised) classes in other languages. Passive classes are used for representing the state of active objects and data passed between active objects. Passive classes are indicated with the keyword **passive**.

```
1  passive class Person {
2    name : string
3    age  : int
4
5    def init(name : string, age : int) {
6      this.name = name;
7      this.age = age;
8    }
9  }
```

In passive classes, all fields are public:

```
1  class Main
2    def main() : void
3      let p = new Person("Dave", 21) in
4        print("Hello {}\n", p.name) -- prints"Hello Dave"
```

4.4 Parametric Classes

Classes can take type parameters. This allows, for example, parameterised pairs
to be implemented:

```
1  passive class Pair<a, b>
2    fst : a
3    snd : b
4    def init(fst_ : a, snd_ : b) : void {
5      this.fst = fst_;
6      this.snd = snd_
7    }
```

This class can be used as follows:

```
1  class Main
2    def main() : void
3      let pair = new Pair<int,string>(65,"a") in
4        print("({},{})\n", pair.fst, pair.snd)
```

Currently, type parameters are unbounded in ENCORE, but this limitation will
be removed in the future.

4.5 Traits and Inheritance

ENCORE is being extended with support for *traits* [14] to be used in place of
standard class-based inheritance. A trait is a composable unit of behaviour that
provides a set of methods and requires a set of fields and methods from any class
that wishes to include it. The exact nature of ENCORE traits has not yet been
decided at time of writing.

A class may be self-contained, which is the case for classes shown so far and
most classes shown in the remainder of this document, or be constructed from
a set of pre-existing traits. The inclusion order of traits is insignificant, and
multiple ways to combine traits are used by the type system to reason about
data races (cf. Sect. 10). Below, the trait Comparator implementation requires
that the including class defines a cmp method, and provides five more high-level
methods all relying on the required method.

```
1 trait Comparator<t>
2   require def cmp(t): int;
3
4   def equal(v:t) : bool
5     this.cmp(v) == 0
6
7   def lessThan(v:t) : bool
8     this.cmp(v) < 0
9   def lessThanOrEqual(v:t) : bool
10    this.cmp(v) <= 0
11
12  def greaterThan(v:t) : bool
13    this.cmp(v) > 0
14  def greaterThanOrEqual(v:t) : bool
15    this.cmp(v) >= 0
```

Traits enable trait-based polymorphism—it is possible, for instance, to write a method that operates on any object whose class includes the Comparator trait:

```
1 def contains(p:person, ps:[Comparator<Person>]) : bool
2   let
3     found = false
4     size = |ps|
5     i = 0
6   in {
7     while not found and i < size
8       {
9         if ps[i].equal(p) then found = true;
10        i = i + 1;
11      }
12    return found;
13  }
```

For more examples of traits, see Sect. 10.

5   Method Calls

Method calls may run asynchronously (returning a future) or synchronously depending primarily on whether the target is active or passive. The complete range of possibilities is given in the following table:

	Synchronous	Asynchronous
Active objects	get o.m()	o.m()
Passive objects	o.m()	—
this (in Active)	this.m()	let that = this in that.m()

Self calls on active objects can be run synchronously—the method called is run immediately—or asynchronously—a future is immediately returned and the invocation is placed in the active object's queue.

Sometimes the result of an asynchronous method call is not required, and savings in time and resources can be gained by not creating the data structure implementing the future. To inform the compiler of this choice, the . in the method call syntax is replaced by a !, as in the following snippet:

```
1 cart ! add_item(item)
```

6 Futures

Method calls on active objects run asynchronously, meaning that the method call is run potentially by a different active object and that the current active object does not wait for the result. Instead of returning a result of the expected type, the method call returns an object called a *future*. If the return type of the method is t, then a value of type **Fut** t is returned to the caller. A future of type **Fut** t is a container that at some point *in the future* will hold a value of type t, typically when some asynchronous computation finishes. When the asynchronous method call finishes, it writes its result to the future, which is said to be *fulfilled*. Futures are considered first class citizens, and can be passed to and returned from methods, and stored in data types. Holding a value of type **Fut** t gives a hint that there is some parallel computation going on to fulfil this future. This view of a future as a handle to a parallel computation is exploited further in Sect. 9.

Several primitive operations are available on futures:

- get: Fut t -> t waits for the future to be fulfilled, blocking the current active object until it is; returns the value stored in the future.
- await: Fut t -> void waits for the future to be fulfilled, without blocking the current active object, thus other methods can run; does not return a value.[1]
- chaining: ~~> : Fut t -> (t -> t') -> Fut t' takes a closure to run on the result when the future is fulfilled; returns another future that will contain the result of running the closure.

These operations will be illustrated using following the classes as a basis. These classes model a service provider that produces a certain product:

```
1 passive class Product { ... }
2 class Service {
3   def provide(): Product {
4     new Product()
5   }
```

The next subsections provide several implementations of clients that call on the service provider, create an instance of class Handle to deal with the result, and pass the result provided by the service provider to the handler.

---

[1] This design should change, so that **await** will become more similar to **get**, but with a different effect on the active object.

6.1   Using the get operation

When the **get** operation is applied to a future, the current active object blocks until the future is fulfilled, and when it has been, the call to **get** returns the value stored in the future.

Consider the following client code.

```
1  class Handler { ... }
2  class Client
3    service : Service
4
5    def run() : void {
6      let fut = service.provide()
7          handler = new Handler()
8      in {
9          handler.handle(get fut);
10         ...
11     }
12   }
```

In method run of Client, the call to service.provide() results in a future of type **Fut** Product (line 6). In line 9, the actual Product object is obtained using a call to **get**. If the future had already been fulfilled, the Product object would be returned immediately. If not, method *and* the active object block, preventing any progress locally until the future is fulfilled.

6.2   Using the await command

One of the problems with calling **get** on a future is that it can result in the entire active object being blocked—sometimes this is desirable to ensure than internal invariants hold, but it can result in costly delays, for example, if the method called involves a time-consuming calculation. During that time, the whole active object can make no progress, which would also block other active objects that need its services.

An alternative, when it makes sense, is to allow the active object to process other messages from its message queue and resume the current method call sometime after the future has been fulfilled. This is exactly what calling **await** on a future allows.

Command **await** applies to a future and waits for it to be fulfilled, blocking the current method call but without blocking the current active object. The call to **await** does not return a value, so a call to **get** is required to get the value. This call to **get** is guaranteed to succeed without blocking.

The following code provides an alternative implementation to the method run from class Client above using **await**:[2]

```
1  def run() : void {
2    let fut = service.provide()
3        handler = new Handler()
```

---

[2] Ideally, this should be: handler.handle(**await** fut). Future versions of ENCORE will support this semantics.

```
4   in {
5       await fut;
6       handler.handle(get fut);
7       ...
8   }
9 }
```

In this code, the call **await fut** on line 5 will block if the future fut is unfulfilled; other methods could run in the same active object between lines 5 and 6. When control returns to this method invocation, execution will resume on line 6 and the call to **get** fut is guaranteed to succeed.

6.3   Using Future Chaining

The final operation of futures is future chaining (fut $\leadsto$ g) [24]. Instead of waiting for the future fut to be fulfilled, as is the case for **get** and **await**, future chaining attaches a closure g to the future to run when the future is fulfilled. Future chaining immediately returns a future that will store the result of applying the closure to the result of the original future.

The terminology comes from the fact that one can add a further closure onto the future returned by future chaining, and add a additional closure onto that, and so forth, creating a chain of computations to run asynchronously. If the code is written in a suitably stylised way (e.g., one of the ways of writing monadic code such as Haskell's do-notation [30]), then the code reads in sequential order—no inversion of control.

Consider the following alternative implementation of the run method from class Client above using future chaining:

```
1 def run() : void {
2   let fut = service.provide()
3       handler = new Handler()
4   in {
5       fut ~~> (\(prod: Producer) -> handler.handle(prod))  -- future chaining
6       ...
7   }
8 }
```

In the example above, the closure defined on line 5 will be executed as soon as the future from service.provide() (line 2) is fulfilled.

A chained closure can run in one of two modes, depending on what is accessed within the closure. If the closure accesses fields or passive objects from the surrounding context, which would create the possibility of race conditions, then it must be run in *attached* mode, meaning that the closure when invoked will be run by the active object that lexically encloses it. The closure in the example above needs to run in attached mode as it accesses the local variable handle. In contrast, a closure that cannot cause race conditions with the surrounding active object can be run in *detached* mode, which means that it can be run independently of the active object. To support the specification of detached closures, the notion of spore [26], which is a closure with a pre-specified environment, can be used (cf. Sect. 7.6). Capabilities (Sect. 10) will also provide means for allowing safe detached closures.

7   Expressions, Statements, and so Forth

ENCORE has many of the language features one expects from a general purpose programming language. Some of these features are described (briefly) in the following subsections.

7.1   Types

ENCORE has a number of built in types. The following table presents these, along with typical literals for each type:

Type	Description	Literals
void	The unit value	()
string	Strings	"hello"
int	Fixed-precision integers	1, -12
uint	Unsigned, fixed-precision integers	42
real	Floating point numbers	1.234, -3.141592
bool	Booleans	true, false
Fut t	Futures of type t	—
Par t	Parallel computations producing type t	—
Stream t	functional streams of type t	—
t -> t'	functions from type t to type t'	\x -> x * 2
[t]	arrays of type t	[1,2,3,6], but not []

The programmer can also introduce two new kinds of types: active class types and passive classes types, both of which can be polymorphic (cf. Sect. 4).

7.2   Expression Sequences

Syntactically, method bodies, while bodies, let bodies, etc. consist of a single expression:

```
1 def single() : void
2   print"a single expression needs no curly braces"
```

In this case, curly braces are optional.

```
1 def curly() : void {
2   print".. but it CAN use them!"
3 }
```

If several expressions need to be sequenced together, this is done by separating them by semicolons and wrapping them in curly braces. The value of a sequence is the value of its last expression. A sequence can be used wherever an expression is expected.

```
1 def multiple() : int {
2     print"multiple";
3     print"expressions";
4     print"are wrapped by { ... }";
5     print"and separated by ';'";
6     2
7 }
```

7.3   Loops

ENCORE has two kinds of loops: **while** and **repeat** loops. A **while** loop takes a boolean loop condition, and evaluates its body expression repeatedly, as long as the loop condition evaluates to true:

```
1 let i = 0 in
2   while i < 5 {
3     print("i={}\n",i);
4     i = i + 1
5   }
```

This prints:

```
i=0
i=1
i=2
i=3
i=4
```

The **repeat** loop is syntax sugar that makes iterating over integers simpler. The following example is equivalent to the **while** loop above:

```
1 repeat i <- 5
2   print("i={}\n",i)
```

In general,

```
1 repeat i <- n
2   expr
```

evaluates expr for values $i = 0, 1, \ldots, n - 1$.

7.4   Arrays

The type of arrays of type T is denoted [T]. An array of length n is created using **new** [T](n). Arrays are indexed starting from 0. Arrays are fixed in size and cannot be dynamically extended or shrunk.

Array elements are accessed using the bracket notation: a[i] accesses the ith element. The length of an array is given by |a|. Arrays can be constructed using the literal notation [1, 2, 1+2].

The following example illustrates the features of arrays:

```
1 class Main
2   def bump(arr: [int]): void
3     repeat i <- |arr|
```

```
4        arr[i] = arr[i] + 1
5
6    def main(): void {
7      let a = [1,2,3] in {
8        this.bump(a);
9        repeat i <- |a|
10         print a[i];
11        let b = new [int](3) in {
12          b[0] = 0;
13          b[1] = a[0];
14          b[2] = 42 - 19;
15        };
16        repeat i <- |b|
17          print b[i];
18      }
19    }
```

The expected output is

```
2
3
4
0
2
23
```

7.5 Formatted Printing

The print statement allows formatted output. It accepts a variable number of parameters. The first parameter is a format string, which has a number of holes marked with {} into which the values of the subsequent parameters are inserted. The number of occurrences of {} must match the number of additional parameters.

The following example illustrates how it works.

```
1  class Main
2    def main() : void {
3      let i = 0 in {
4        while i < 5 {
5          i = i+1;
6          print("{} * {} = {}\n", i, i, i*i);
7        }
8      }
9    }
```

The output is:

```
$ ./ex_printing.enc
1 * 1 = 1
2 * 2 = 4
3 * 3 = 9
4 * 4 = 16
5 * 5 = 25
```

7.6   Anonymous Functions

In ENCORE, an anonymous function is written as follows:

```
1  \(i : int) -> 10 * i
```

This function multiplies its input *i* by 10.

The backslash \ (syntax borrowed from Haskell, resembling a lambda) is followed by a comma separated list of parameter declarations, an arrow -> and an expression, the function body. The return type does not need to be declared as it is always inferred from the body of the lambda.

In the example below, the anonymous function is assigned to the variable tentimes and then later applied—it could also be applied directly.

```
1  let tentimes = \(i : int) -> 10 * i in
2      print(tentimes(10)) -- prints 100
```

Anonymous functions are first-class citizens and can be passed as arguments, assigned to variables and returned from methods/functions. Types of functions are declared by specifying its arguments types, an arrow ->, and the return type. For example, the type of the function above is **int ->int**. Multi-argument functions have types such as **(int, string)->bool**.

The following example shows how to write a higher-order function update that takes a function f of type **int->int**, an array of **int**'s and applies the function f to the elements of the array data, updating the array in-place.

```
1  def update(f: int -> int, data: [int]): void {
2    repeat i <- |data|
3        data[i] = f(data[i]);
4  }
5
6  class Main
7    def main(): void {
8      let xs = [2,3,4,1] in
9        update(\(data: int) -> data + 1, xs)
10   }
```

Closures as specified above can capture variables appearing in their surrounding lexical context. If a closure is run outside of the context in which it is defined, then data races can occur. A variation on closures exists that helps avoid this problem.

ENCORE provides a special kind of anonymous function called a *spore* [26]. A spore must explicitly specify the elements from its surrounding context that are captured in the spore. The captured elements can then, more explicitly, be controlled using types, locks or cloning to ensure that the resulting closure can be run outside of the context in which the spore is defined. Spores have an environment section binding the free variables of the sport to values from the surrounding context, and a closure, which can access only those free variables and its parameters.

```
1  class Provider
2    service: Service
3
4    def provide(): Data -> Product {
```

```
5    spore {
6      let x = clone this.service in -- set up environment for closure
7      \(y: Data) -> x.produce(y) -- the closure
8    }
9  }
```

In this code snippet, the only variables in scope in the closure body are x and y. The field service, which would normally be visible within the closure (in Scala or in Java if it were final), is *not* accessible. It is made accessible (actually, a clone of its contents), via variable x in the environment section of the spore.

7.7 Polymorphism and Type Inference

At the time of writing, ENCORE offers some support for polymorphic classes, methods and functions. Polymorphism in ENCORE syntactically resembles other well-established OOP languages, such as Java. Type variables in polymorphic classes, methods and/or functions must be written using lower case.

The following example shows how to write a polymorphic list:

```
1  passive class List<t>
2    data: t
3    next: List<t>
4
5    def init(data: t): void
6      this.data = data
7
8    def append(data: t): void {
9      let next_item = new List<t>(this.data) in {
10       next_item.next = this.next;
11       this.data = data;
12       this.next = next_item;
13     }
14   }
15   -- other methods
16
17 class Main
18   def main(): void {
19     let l = new List<int> in {
20       l.append(3);
21       l.append(4);
22     }
23   }
```

7.8 Module System

Currently, ENCORE supports a rudimentary module system. The keyword **import** followed by the name of a module imports the corresponding module. The name of the module must match the name of the file, excluding the .enc suffix. The compiler looks for the corresponding module in the current directory plus any directories specified using the -I pathlist compiler flag.

Assume that a library module `Lib.enc` contains the following code:

```
1  class Foo
2    def boo(): void {
3      print"^-^"
4    }
```

This module can be imported using `import Lib` as illustrated in the following (file `Bar.enc`):

```
1  import Lib
2
3  class Main
4    def main(): void {
5      let
6        f = new Foo
7      in
8        f.boo()
9    }
```

Here `Bar.enc` imports module `Lib` and can thus access the class `Foo`.

Currently, the module system has no notion of namespaces, so all imported objects needs to have unique names. There is also no support for cyclic imports and qualified imports, so it is up to the programmer to ensure that each file is only imported once.

7.9 Suspending Execution

The **suspend** command supports cooperative multitasking. It suspends the currently running method invocation on an active object and schedules the invocation to be resumed after all messages in the queue have been processed.

```
1  class Pi
2    def calculate_digits(digits: int): double {
3      -- perform initial calculations
4      ...
5      suspend;
6      -- continue performing more calculations
7      ...
8    }
9
10 class Main
11   def main(): void {
12     let pi = new Pi() in {
13       pi.calculate_decimals(100000000000);
14     }
15   }
```

The example computes a large number of digits of π. The method `calculate_digits` calls **suspend** to allow other method calls to run on the `Pi` active object. This is achieved by suspending the execution of the current method call, placing a new message in its message queue, and then releasing control. The message placed in the queue is the continuation of the suspended method invocation, which in this case will resume the suspended method invocation at line 6.

7.10   Embedding of C Code

ENCORE supports the embedding of C code. This is useful for wrapping C libraries to import into the generated C code and for experimenting with implementation ideas before incorporating them into the language, code generator, or run-time. Two modes are supported: top-level embed blocks and embedded expressions.

Note that we do not advocate the extensive use of **embed**. Code using **embed** is quite likely to break with future updates to the language.

Top-Level Embed Blocks. Each file can contain at most one top-level **embed** block, which has to be placed before the first class definition in the file. This **embed** block consists of a header section and an implementation section, as in the following example:

```
1 embed
2   int64_t sq(int64_t);
3 body
4   int64_t sq(int64_t n) {
5     return n*n;
6   }
7 end
```

The header section will end up in a header file that all class implementations will include. The implementation section will end up in a separate C file. The sq function declaration must be included in the header section, otherwise the definitions in the **body** section would not be accessible in the generated C code.

Embedded Expressions. An **embed** block can appear anywhere where an expression can occur. The syntax is:

```
1 embed encore-type C-code end
```

When embedding an expression, the programmer needs to assign an encore type to the expression. ENCORE will assume that this type is correct. The value of an embedded expression is the value of the last C-statement in the embedded code.

ENCORE variables can be accessed from within an **embed** block by wrapping them with #{ }. For instance, local variable x in ENCORE code is accessed using #{x} in the embedded C. Accessing fields of the current object is achieved using C's arrow operator. For instance, **this**->foo accesses the field **this**.foo.

The following example builds upon the top-level embed block above:

```
1 class Main
2   def main() : void {
3     let x = 2 in
4       print(embed int sq(#{x}); end)
5   }
```

The embedded expression in this example promises to return an int. It calls the C-function sq on the local ENCORE variable x.

Embedding C Values as Abstract Data Types. The following pattern allows C values to be embedded into ENCORE code and treated as an abstract type, in a sense, where the only operations that can manipulate the C values are implemented in other embedded blocks. In the following code example, a type D is created with no methods or fields. Values of this type cannot be manipulated in ENCORE code, only passed around and manipulated by the corresponding C code.

```
 1  passive class D
 2
 3  passive class LogArray
 4    size:int
 5    slots:D
 6    def init(size:int) : void
 7      embed void -- initialise element of type D
 8        this->slots = pony_alloc(size * sizeof(void*));
 9        for (int i = 0; i < size; ++i) ((pony_actor_t**)this->slots)[i] = NULL;
10        this->size = size;
11      end
12    def write(i:int, v:LogEntry) : void
13      embed void -- modify element of type D
14      ((void **)this->slots)[i] = v;
15      end
16    def read(i:int) : LogEntry
17      embed LogEntry --- read element of type D
18      ((void **)this->slots)[i];
19      end
20    def size() : int
21      this.size
```

Mapping ENCORE Types to C Types. The following table documents how ENCORE's types are mapped to C types. This information is useful when writing embedded C code, though ultimately having some detailed knowledge of how ENCORE compiles to C will be required to do anything advanced.

ENCORE type	C type
string	(char *)
real	double
int	int64_t
uint	uint64_t
bool	int64_t
⟨an active class type⟩	(encore_actor_t *)
⟨a passive class type⟩	(CLASSNAME_data *)
⟨a type parameter⟩	(void *)

8   Streams

A stream in ENCORE is an immutable sequence of values produced asynchronously. Streams are abstract types, but metaphorically, the type **Stream** a can be thought of as the Haskell type:

```
1  type Stream a = Fut (Maybe (St a))
2  data St a = St a (Stream a)
```

That is, a stream is essentially a future, because at the time the stream is produced it is unknown what its contents will be. When the next part of contents are known, it will correspond to either the end of the stream (Nothing in Haskell) or essentially a pair (Just (St e s)) consisting of an element e and the rest of the stream s.

In ENCORE this metaphor is realised, imperfectly, by making the following operations available for the consumer of a stream:

- **get: Stream** a -> a — gets the head element of the (non-empty) stream, blocking if it is not available.
- **getNext: Stream** a -> **Stream** a — returns the tail of the (non-empty) stream. A non-destructive operator.
- **eos: Stream** a -> Bool — checks whether the stream is empty.

Streams are produced within special **stream** methods. Calling such methods results immediately in a handle to the stream (of type **Stream** a). Within such a method, the command **yield** becomes available to produce values on the stream. **yield** takes a single expression as an argument and places the corresponding value on the stream being produced. When the stream method finishes, stream production finishes and the end of the stream marker is placed in the stream.

The following code illustrate an example stream producer that produces a stream whose elements are of type **int**:

```
1  class IntSeq
2    stream start(fr : int, to : int) : int {
3      while fr <= to {
4        yield fr;
5        fr = fr+1
6      };
7    }
```

The following code gives an example stream consumer that processes a stream stored in variable str of type **Stream int**.

```
1  class Main
2    def main() : void
3      let
4        lst = 0
5        str = (new IntSeq).start(1,1000000)
6      in {
7        while not eos str {
8          lst = get str;
```

```
 9        str = getNext str;
10      };
11      print lst
12    }
```

Notice that the variable str is explicitly updated with a reference to the tail of the stream by calling **getNext**, as **getNext** returns a reference to the tail, rather than updating the object in str in place—streams are immutable, not mutable.

9 Parallel Combinators

ENCORE offers preliminary support for parallel types, essentially an abstraction of parallel collections, and parallel combinators that operate on them. The combinators can be used to build pipelines of parallel computations that integrate well with active object-based parallelism.

9.1 Parallel Types

The key ingredient is the parallel type **Par** t, which can be thought of as a handle to a collection of parallel computations that will eventually produce zero or more values of type t—for convenience we will call such an expression a *parallel collection*. (Contrast with parallel collections that are based on a collection of elements of type t manipulated using parallel operations [31].) Values of **Par** t type are first class, thus the handle can be passed around, manipulated and stored in fields of objects.

Parallel types are analogous to future types in a certain sense: an element of type **Fut** t can be thought of as a handle to a single asynchronous (possibly parallel) computation resulting in a single value of type t; similarly, an element of type **Par** t can be thought of as a handle to a parallel computation resulting in multiple values of type t. Pushing the analogy further, **Par** t can be thought of as a "list" of elements of type **Fut** t: thus, **Par** t $\approx$ [**Fut** t].

Values of type **Par** t are assumed to be ordered, thus ultimately a sequence of values as in the analogy above, though the order in which the values are produced is unspecified. Key operations on parallel collections typically depend neither on the order the elements appear in the structure nor the order in which they are produced.[3]

9.2 A Collection of Combinators

The operations on parallel types are called parallel combinators. These adapt functionality from programming languages such as Orc [22] and Haskell [30] to express a range of high-level typed coordination patterns, parallel dataflow pipelines, speculative evaluation and pruning, and low-level data parallel computations.

---

[3] An alternative version of **Par** t is possible where the order in the collection is not preserved. This will be considered in more detail in future experiments.

The following are a representative collection of operations on parallel types.[4]
Note that all operations are functional.

- `empty`: **Par** t. A parallel collection with no elements.
- `par`: (**Par** t, **Par** t) ->**Par** t. The expression par(c, d) runs c and d in parallel and results in the values produced by c followed (spatially, but not temporally) by the values produced by d.
- `pbind`: (**Par** t, t ->**Par** t') -> Par t'. The expression pbind(c,f) applies the function f to all values produced by c. The resulting nested parallel collection (of type **Par** (**Par** t')) is flattened into a single collection (of type **Par** t'), preserving the order among elements at both levels.
- `pmap`: (t -> t', Par t) -> Par t' is a parallel map. The expression pmap(f,c) applies the function f to each element of parallel collection c in parallel resulting in a new parallel collection.
- `filter`: (t ->**bool**, **Par** t) -> **Par** t filters elements. The expression filter (p, c) removes from c elements that do not satisfy the predicate p.
- `select`: **Par** t -> Maybe t returns the first available result from the parallel type wrapped in tag Just, or Nothing if it has no results.[5]
- `selectAndKill`: **Par** t -> Maybe t is similar to select except that it also kills all other parallel computations in its argument after the first value has been found.
- `prune`: (**Fut** (Maybe t) ->**Par** t', Par t) -> Par t'. The expression prune(f, c) creates a future that will hold the result of selectAndKill(c) and passes this to f. This computation in run in parallel with c. The first result of c is passed to f (via the future), after which c is terminated.
- `otherwise`: (**Par** t, () ->**Par** t) ->**Par** t. The expression otherwise (c, f) evaluates c until it is known whether it will be empty or non-empty. If it is not empty, return c, otherwise return f().

A key omission from this list is any operation that actually treats a parallel collection in a sequential fashion. For instance, getting the first (leftmost) element is not possible. This limitation is in place to discourage sequential programming with parallel types.

9.3   From Sequential to Parallel Types

A number of functions lift sequential types to parallel types to initiate parallel computation or dataflow.

---

[4] As work in parallel types and combinators is work in progress, this list is likely to change and grow.

[5] Relies on Maybe data type: in Haskell syntax data Maybe a = Nothing | Just a. Data types are at present being implemented. An alternative to Maybe is to use **Par** restricted to empty and singleton collections. With this encoding, the constructors for Maybe become Nothing = empty, Just a = liftv a (from Sect. 9.3), and the destructor, maybe :: b -> (a -> b) -> Maybe a -> b in Haskell, is defined in ENCORE as **def** maybe(c: b, f: a->b, x: Maybe a) = otherwise(pmap(f, x), \() -> c).

- `liftv :: t -> Par t` converts a value to a singleton parallel collection.
- `liftf :: Fut t -> Par t` converts a future to a singleton parallel collection (following the `Par t` ≈ `[Fut t]` analogy above).
- `each :: [t] -> Par t` converts an array to a parallel collection.

One planned extension to provide a better integration between active objects and parallel collections is to allow fields to directly store collections but not as a data type but, effectively, as a parallel type. Then applying parallel operations would be more immediate.

9.4   ... and Back Again

A number of functions are available for getting values out of parallel types. Here is a sample:

- `select :: Par t -> Maybe t`, described in Sect. 9.2, provides a way of getting a single element from a collection (if present).
- `sum :: Par Int -> Int` and other fold/reduce-like functions provide operations such as summing the collection of integers.
- `sync :: Par t -> [t]` synchronises the parallel computation and produces a sequential array of the results.
- `wsync :: Par t -> Fut [t]` same as `sync`, but instead creates a computation to do the synchronisation and returns a future to that computation.

9.5   Example

The following code illustrates parallel types and combinators. It computes the total sum of all bank accounts in a bank that contain more than 10,000 euros.

The program starts by converting the sequential array of customers into a parallel collection. From this point on it applies parallel combinators to get the accounts, then the balances for these accounts, and to filter the corresponding values. The program finishes by computing a sum of the balances, thereby moving from the parallel setting back to the sequential one.

```
1  import party
2
3  class Main
4    bank : Bank
5    def main(): void {
6      let
7        customers = each(bank.get_customers()) -- get customers objects
8        balances =
9          filter(\(x: int) -> { x > 10000 }, -- filter accounts
10            pmap(\(x: Account) -> x.get_balance()), -- get all balances
11              pbind(customers,
12                    \(x : Customer) -> x. get_accounts())) -- get all accounts
13      in
14        print("Total: {}\n", sum(balances))
15    }
```

9.6 Implementation

At present, parallel types and combinators are implemented in ENCORE as a library and the implementation does not deliver the desired performance. In the future, `Par` t will be implemented as an abstract type to give the compiler room to optimise how programs using parallel combinators are translated into C. Beyond getting the implementation efficient, a key research challenge that remains to be addressed is achieving safe interaction between the parallel combinators and the existing active object model using capabilities.

10 Capabilities

The single thread of control abstraction given by active objects enables sequential reasoning inside active objects. This simplifies programming as there is no interleaving of operations during critical sections of a program. However, unless proper encapsulation of passive objects is in place, mutable objects might be shared across threads, effectively destroying the single thread of control.

A simple solution to this problem is to enforce deep copying of objects when passing them between active objects, but this can increase the cost of message sending. (This is the solution adopted in the formal semantics of ENCORE presented in Sect. 12.) Copying is, however, not ideal as it eliminates cases of benign sharing of data between active objects, such as when the shared data is immutable. Furthermore, with the creation of parallel computation inside active objects using the parallel combinators of Sect. 9, more fine-grained ways of orchestrating access to data local to an active object is required to avoid race conditions.

These are the problems addressed by the *capability type system* in ENCORE.[6]

10.1 Capabilities for Controlling of Sharing

A *capability* is a token that governs access to a certain resource [27]. In an attempt to re-think the access control mechanisms of object-oriented programming systems, ENCORE uses capabilities *in place of* references and the resources they govern access to are objects, and often entire aggregates. In contrast to how references normally behave in object-oriented programming languages, capabilities impose principles on how and when several capabilities may govern access to a common resource. As a consequence, different capabilities impose different means of alias control, which is statically enforce at compile-time.

In ENCORE capabilities are constructed from *traits*, the units of reuse from which classes can be built (cf. Sect. 4.5). Together with a *kind*, each trait forms a capability, from which composite capabilities can be constructed. A capability provides an interface, essentially the methods of the corresponding trait. The capability's kind controls how this interface can be accessed with respect to

---

[6] Note that at the time of writing, the capability system has not been fully implemented.

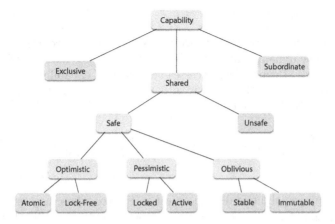

Fig. 2. The hierarchy of capabilities. Leaf nodes denote concrete capabilities, non-leaves categorise.

avoiding data-races. Capabilities are combined to form classes, just as traits do, which means that the range of capabilities are specified at class definition, rather than at object creation time.

From an object's type, it is immediately visible whether it is *exclusive* to a single logical thread of control (which trivially implies that accesses to the object are not subject to data races), *shared* between multiple logical threads (in which case freedom from data races must be guaranteed by some concurrency control mechanism), or *subordinate* to some other object which protects it from data races—this is the default capability of a passive class in ENCORE. The *active* capability is the capability kind of active classes in ENCORE. Figure 2 shows the different capabilities considered, which will be discussed below.[7]

10.2 Exclusive Capabilities

Exclusive capabilities are exclusive to a single thread of control. exclusive capabilities implement a form of *external uniqueness* [12] where a single pointer is guaranteed to be the only *external* pointer to an entire aggregate. The uniqueness of the external variable is preserved by destructive reads, that is, the variable must be nullified when read unless it can be guaranteed that the two aliases are not visible to any executing threads at the same time.

Exclusive capabilities greatly simplify ownership transfer—passing an exclusive capability as an argument to a method on another active object requires the nullification of the source variable, which means that all aliases at the source to the entire aggregate are dead and that receiver has sole access to the transferred object.

---

[7] ENCORE may eventually *not* include all kinds of capabilities presented here, this is a matter under consideration.

10.3   Shared Capabilities

A shared capability expresses that the object is shared and, in contrast to exclusive capabilities, some dynamic means is required to guarantee data-race freedom.

The semantics of concurrent accesses via a shared capability is governed by the sharing kind. The list below overviews the semantics of concurrent accesses of safe (first six) and unsafe capabilities.

Active. Active capabilities denote ENCORE's active classes. They guarantee race-freedom through dynamic, pessimistic concurrency control by serialising all its inside computation.

Atomic. Atomic capabilities denote object references whose access is managed by a transaction. Concurrent operations either commit or rollback in a standard fashion. From a race-freedom perspective, atomic capabilities can be freely shared across ENCORE active objects.

An interesting question arises when considering the interaction between transactions and asynchronous message passing: can asynchronous messages "escape" a transaction? Since asynchronous messages are processed by other logical threads of control, they may be considered a side-effect that is impossible to roll-back. Some ways of resolving this conundrum are:

1. Forbidding asynchronous message sends inside transactions. Problem: *Restricts expressivity.*
2. Delaying the delivery of asynchronous message sends to commit-time of a transaction. Problem: *Reduces throughput/increases latency.*
3. Accepting this problem and leaving it up to programmer's to ensure the correctness of their programs when transactions are mixed with asynchronous message passing. Problem: *Not safe.*

Immutable. Immutable capabilities describe data that is always safely accessible without concurrency control. Immutability is "deep", meaning that state can be observably modified through an immutable reference, though a method in an immutable object can mutate state created within the method or of its arguments. Immutable capabilities can be freely shared across ENCORE active objects without any need for copying.

Locked. Each operation via a locked capability requires prior acquisition of a lock specific for the resource. The lock can be reentrant (analogous to a synchronised method in Java), a readers-writer lock, etc. depending on desired semantics.

LockFree. The implementations of behaviours for this capability must follow a certain protocol for coordinating updates in a lock-free manner. Lock-free programming is famously subtle, because invariants must be maintained at all times, not just at select commit-points. As part of the work on ENCORE, we are implementing a type system that enforces such protocol usage on lock-free capabilities [8].

Stable. Stable capabilities present an immutable view of otherwise mutable state. There are several different possible semantics for stable capabilities: *read-only references*—capability cannot be used to modify state, but it may witness changes occurring elsewhere; *fractional permissions*—if a stable capability is available, no mutable alias to an overlapping state will be possible, thereby avoiding read-write races; or *readers-writer locks*—a static guarantee that a readers-writer lock is in place and used correctly.

Unsafe. As the name suggest, unsafe capabilities come with no guarantees with respect to data races. Allowing unsafe capabilities is optional, but they may be useful to give a type to embedded C code.

10.4 Subordinate Capabilities

A subordinate capability is a capability that is dominated by an exclusive or shared capability, which means that the dominating capability controls access to the subordinate. In ENCORE, passive objects are subordinate capabilities by default, meaning they are encapsulated by their enclosing active object. This corresponds to the fact that there can be no "free-standing" passive objects in ENCORE, they all live on the local heap of some active object.

Encapsulation of subordinate objects is achieved by disallowing them to be passed to non-subordinate objects. A subordinate capability is in this respect similar to the owner annotation from ownership types [10,38].

Some notion of borrowing can be used to safely pass subordinate objects around under some conditions [13].

10.5 Polymorphic Concurrency Control

Capabilities allow for polymorphic concurrency control through the abstract capabilities *shared, safe, optimistic, pessimistic* and *oblivious*. This allows a library writer to request that a value is protected from data races by some means, but not specify those means explicitly. For example:

```
1 def transfer(from:safe Account, to:safe Account, amount:int) : void
2   to.deposit(from.withdraw(amount))
```

This expresses that the calls on `from` and `to` are safe from a concurrency stand-point. However, whether this arises from the accounts using locks, transactions or immutability is not relevant here.

Accesses through safe capabilities are interesting because the semantics of different forms of concurrency control requires a small modicum of extra work at run-time. For example, if `from` is active, then `from.withdraw()` should (implicitly) be turned into **get** `from.withdraw()`, or if we are *inside* an atomic capability and `to` is a locked capability, then the transfer transaction should be extended to also include `to.deposit()`, and committing the transaction involves being able to grab the lock on `to` and release it once the transaction's log has been synchronised with the object.

The exact semantics of the combinations are currently being worked out.

10.6   Composing Capabilities

A single capability is a trait plus a mode annotation. Mode annotations are the labels in Fig. 2. Leaves denote concrete modes, i.e., modes that can be used in the definition of a capability or class. Remaining annotations such as safe, pessimistic etc. are valid only in types to abstract over concrete annotations, or combinations of concrete annotations.

Capabilities can be composed in three different ways: *conjunction* $C_1 \otimes C_2$, *disjunction* $C_1 \oplus C_2$, and *co-encapsulation* $C_1 \langle C_2 \rangle$.

A conjunction or a disjunction of two capabilities C_1 and C_2 creates a composite capability with the union of the methods of C_1 and C_2. In the case of a disjunction, C_1 and C_2 may share state without concurrency control. As a result, the same guard (whether it is linearity, thread-locality, a lock, etc.) will preserve exclusivity of the entire composite. In the case of a conjunction, C_1 and C_2 must not share state, except for state that is under concurrency control. For example, they may share a common field holding a shared capability, as long as neither capability can write the field. The conjunction of C_1 and C_2, $C_1 \otimes C_2$, can be unpacked into its two sub-capabilities C_1 and C_2, creating two aliases to the same object that can be used without regard for the other.

In contrast to conjunction and disjunction, co-encapsulation denotes a nested composition, where one capability is buried inside the other, denoted $C_1 \langle C_2 \rangle$. The methods of the composite $C_1 \langle C_2 \rangle$ are precisely those of C_1, but by exposing the nested type C_2 in the interface of the composite capability, additional operations on the type-level become available. Co-encapsulation is useful to preserve linearity of nested capabilities. For example, unless C_3 is exclusive, the capability $C_3 \langle C_1 \otimes C_2 \rangle$ can be turned into $C_3 \langle C_1 \rangle \otimes C_3 \langle C_2 \rangle$ which introduces aliases to C_3 but in a way that only disjoint parts of the nested capability can be reached.

Capabilities of different kinds may be used in disjunctions and conjunctions. A capability with at least one exclusive component must be treated linearly to guarantee race-freedom of its data. A capability with at least one subordinate component will be contained inside its enclosing class. There are vast possibilities to create compositions of capabilities, and we are currently investigating their possible uses and interpretations. For example, combinations of active and exclusive capabilities allow operating on an active object as it if was a passive object until the exclusive capability is lost, after which the active object can be freely shared. This gives powerful control over the initialisation phase of an object. As another example, conjunctions of active capabilities could be used to express active objects which are able to process messages in parallel.

10.7   Implementing a Parallel Operation on Disjoint Parts of Shared State

Figures 3, 4, 5, and 6 show how the capabilities can be used to construct a simple linked list data structure of exclusive pairs, which is subsequently "unpacked" into two (logical) immutable lists of disjoint cells which are passed to different objects, and later re-packed into a single mutable list of pairs again. Support for unstructured packing and unpacking is important in ENCORE as communication

```
1  -- declares a capability through a trait
2  trait Cell {
3    -- requirement of including class
4    require var value : int
5
6    def get() : int
7      value
8
9    def set(v : int) : void
10     value = v
11 }
12
13 // constructs class from two exclusive capabilities
14 passive class WeakPair = exclusive Cell ⊗ exclusive Cell' {
15   var value : int -- mutable field
16   var value' : int
17 }
18
19 passive class StrongPair = exclusive Cell ⊕ exclusive Cell' {
20   var value : int
21   var value' : int
22 }
```

Fig. 3. Two different implementations of pairs from the same building blocks. WeakPair uses conjunction ($\otimes$) and StrongPair disjunction ($\oplus$).

across active objects has a more flexible control flow than calls on a single stack, or fork-join style parallelism.

The **trait** keyword introduces a new trait which requires the presence of zero or more fields in any class that includes it. Figure 3 illustrates a trait Cell that requires a mutable field value in any including class.

The compositions of cells into WeakPair and StrongPair have different reuse stories for the Cell trait. The cells of a WeakPair may be independently updated by different threads whereas the cells of a StrongPair always belong to the same thread and are accessed together.

For simplicity, we employ a prime notation renaming scheme for traits to avoid name clashes when a single trait is included more than once.

Figure 4 shows how three capabilities construct a singly linked list. The links in the list are subordinate objects, and the elements in the list are typed by some exclusive parameter P. The capabilities of the List class are Add, Del and Get. The first two are exclusive and the last is stable.

The Add and Del capabilities can add and remove exclusive P objects from the list. (Since these objects are exclusive, looking them up, removes them from the list to maintain linear access.) Since Add and Del share the same field first with the same type, they are not safe to use separately in parallel, so their combination must be a disjunction. If they had been, for example, locked capabilities instead, they would have protected their internals dynamically, so in this case, a conjunction would be allowed.

```
1 passive class List<exclusive P> =
2   exclusive Add<P> ⊕ exclusive Del<P> ⊕ stable Get<P> {
3   var first : Link<P>; // subordinate, so strongly encapsulated
4 }
5
6 trait Add<exclusive P> {
7   require var first : Link<P>;
8
9   def append(v : P) : void {
10     var tmp : Link = new Link<P>();
11     tmp.setNext(first);
12     tmp.setValue(consume v); // destructive read
13   }
14 }
15
16 trait Del<exclusive P> {
17   require var first : Link<P>;
18
19   def remove(i : int) : P {
20     var tmp : Link<P> = first;
21     while (i > 0) {
22       i = i-1;
23       tmp = tmp.getNext();
24     };
25     var v : P = tmp.getValue();
26     ... // code for removing link omitted for brevity
27     return consume v;
28   }
29 }
30
31 trait Get<exclusive P> {
32   // non-assignable ''final'' field
33   require val first : Iterator<P>;
34
35   def lookup(i : int) : S(P) { // stack-bound return
36     var tmp : S(Iterator<P>) = first;
37     while (i > 0) {
38       i = i-1;
39       tmp = tmp.next();
40     };
41     return tmp.value();
42   }
43 }
```

Fig. 4. A list class. P above is a type parameter which allows deep unpacking of the object.

Linearity of exclusive capabilities is maintained by an explicit destructive read, the keyword consumes. The expression **consume** x returns the value of x, and updates x with **null**, logically in one atomic step.

The Get capability overlaps with the others, but the requirement on the field first is different: it considers the field immutable and its type stable through

```
1  passive class Link<exclusive P> = Node<P> ⊕ Iterator<P> {
2    var elmt : P;
3    var next : Link<P>;
4  }
5
6  trait Node<exclusive P> {
7    require var elmt : P;
8    require var next : Link<P>;
9
10   def setNext(n : Link<P>) : void { next = n; }
11
12   def setValue(e : P) : void { elmt = consume e; }
13
14   def getNext() : Link<P> { return next; }
15
16   def getValue() : P { return consume elmt; }
17 }
18
19 trait Iterator<exclusive P> {
20   require val elmt : P;
21   require val next : Iterator<P>;
22
23   def next() : S(Iterator<P>) { return next; }
24
25   def value() : S(P) { return elmt; }
26 }
```

Fig. 5. Definition of the link class with a stable iterator capability to support non-destructive parallel iteration over elements in the list.

the Iterator capability (cf. Fig. 5). As the Get capability's state overlaps with the other capabilities, their composition must be in the form of a disjunction.

Inside the Get trait, the list will not change—the first field cannot be reassigned and the Iterator type does not allow changes to the chain of links. The Get trait however is able to perform *reverse borrowing*, which means it is allowed to read exclusive capabilities on the heap *non-destructively* and return them, as long as they remain stack bound. The stack-bound reference is marked by a type wrapper, such as S(P).

The link capabilities used to construct the list are shown in Fig. 5; they are analogous to the capabilities in List, and are included for completeness.

Finally, Fig. 6 shows the code for unpacking a list of WeakPairs into two logical, stable lists of cells that can be operated on in parallel. The stable capability allows multiple (in this case two) active objects to share part of the list structure with a promise that the list will not change while they are looking.

On a call to start() on a Worker, the list of WeakPairs is split into two in two steps. Step one (line 10–11) turns the List disjunction into a conjunction by *jailing* the Add and Del components which prevents their use until the list is reassembled again on line 35. Step two (line 12) turns the iterator into two by unpacking the pair into two cells.

```
1  class Worker
2    var data : List<WeakPair>
3    var one : J(Add<WeakPair> ⊕ Del<WeakPair>)
4    var two : J(Add<WeakPair> ⊕ Del<WeakPair>) ⊗ Get<Cell>
5    val other : Worker
6    var sum : int
7
8    def start() : void
9      let
10       var stash : J(Add<WeakPair> ⊕ Del<WeakPair>),
11           iter : Get<WeakPair> = consume this.data
12       var fst_i : Get<Cell>, snd_i : Get<Cell> = consume iter
13     in {
14       this.stash = stash; -- keep this part for later
15       this ! do_work(consume fst_i, this);
16       other ! do_work(consume snd_i, this);
17     }
18
19   def do_work(w : Get<Cell>, s : Worker) : void
20     let
21       sum = 0
22     in {
23       repeat i <- w.length()
24         sum = sum + w.get(i).get();
25       s ! done(sum);
26       s ! reassemble(w);
27     }
28
29   def done(sum:int) : void
30     this.sum = this.sum + sum
31
32   def reassemble(part : Get<Cell>) : void
33     if this.two == null
34     then this.two = this.one + part
35     else this.data = this.two + part
```

Fig. 6. Parallel operations on a single list of pairs using unpack and re-packing. Note that this code would not type check for **var** a: List<StrongPair> as StrongPair is built from a disjunction that does not allow unpacking.

The jail construct is used to temporarily render part of a disjunction inaccessible. In the example, Add<WeakPair> ⊕ Del<WeakPair> ⊕ Get<Cell> is turned into J(Add<WeakPair> ⊕ Del<WeakPair>) ⊗ Get<Cell>. The latter type allows unpacking the exclusive reference into two, but since one cannot be used while it is jailed, the exclusivity of the referenced object is preserved.

The lists of cells are passed to two workers (itself and other) that perform some work, before passing the data back for reassembly (line 32).

11 Examples

A good way to get a grip on a new programming language is to study how it is applied in larger programs. To this end, three ENCORE programs, implementing a thread ring (Sect. 11.1), a parallel prime sieve (Sect. 11.2), and a graph generator following the preferential attachment algorithm (Sect. 11.3), are presented.

11.1   Example: Thread Ring

Thread Ring is a benchmark for exercising the message passing and scheduling logic of parallel/concurrent programming languages. The program is completely sequential, but deceptively parallel. The corresponding ENCORE program (Fig. 7) creates 503 active objects, links them forming a ring and passes a message containing the remaining number of hops to be performed from one active object to the next. When an active object receives a message containing 0, it prints its own id and the program finishes.

```
1  class Main
2    def main(): void {
3      let
4        index = 1
5        first = new Worker(index)
6        next = null : Worker
7        nhops = 50*1000*1000
8        ring_size = 503
9        current = first
10     in {
11       while (index < ring_size) {
12         index = index + 1;
13         next = new Worker(index);
14         current!setNext(next);
15         current = next;
16       };
17       current!setNext(first);
18       first!run(nhops);
19     }
20   }
21
22 class Worker
23   id : int
24   next : Worker
25
26   def init(id : int): void
27     this.id = id
28
29   def setNext(next: Worker): void
30     this.next = next
31
32   def run(hops : int): void
33     if (hops > 0) then
34       this.next!run(hops-1)
35     else
36       print(this.id)
```

Fig. 7. Thread ring example

In this example, the active objects forming the ring are represented by the class `Worker`, which has field `id` for worker's id and `next` for the next active object in the ring. Method `init` is the constructor and the ring is set up using method `setNext`. The method `run` receives the number of remaining hops, checks whether this is larger than 0. If it is, it sends an asynchronous message to the next active object with the number of remaining hops decrement by 1. Otherwise, the active object has finished and prints its id.

11.2 Example: Prime Sieve

This example considers an implementation of the parallel Sieve of Eratosthenes in ENCORE. Recall that the Sieve works by filtering out all non-trivial multiples of 2, 3, 5, etc., thereby revealing the next prime, which is then used for further filtering. The parallelisation is straightforward: one active object finds all primes in $\sqrt{N}$ and uses M filter objects to cancel out all non-primes in (chunks of) the interval $[\sqrt{N}, N]$. An overview of the program is found in Fig. 8 and the code is spread over Figs. 9, 10, 11 and 12. Which each filter object finally receives a "done" message, they scan their ranges for remaining (prime) numbers and report these to a special reporter object that keeps a tally of the total number of primes found.

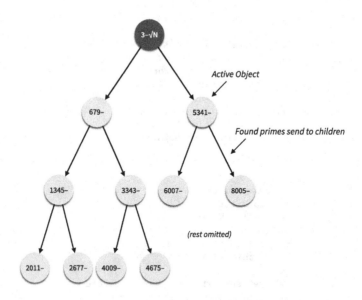

Fig. 8. Overview of the parallel prime sieve. The root object finds all primes in $[2, \sqrt{N}]$ and broadcasts these to filter objects that cancel all multiples of these in some ranges. When a filter receives the final "done" message, it will scan its range for remaining (prime) numbers and report these to an object that keeps a tally.

The listing of the prime sieve program starts by importing libraries. The most important library component is a bit vector data type, implemented as a

```
 1  import lib
 2
 3  class Reporter
 4    primes:int
 5    candidates:int
 6
 7    def init(c:int) : void
 8      this.candidates = c
 9
10    def report(p:int, c:int) : void {
11
12      this.candidates = this.candidates - c;
13      this.primes = this.primes + p;
14
15      if this.candidates == 0 then print this.primes;
16    }
```

Fig. 9. Prime Sieve (a). The Reporter class collects the reports from all filter active objects and summarises the (number of) primes found.

thin ENCORE wrapper around a few lines of C to flip individual bits in a vector. Figure 9 shows the class of the reporter object that collects and aggregates the results of all filter active objects. When it is created, is it told how many candidates are considered (e.g., all the primes in the first 1 billion natural numbers), and as every filter reports in, it reports the number of primes found in the number of candidates considered.

The main logic of the program happens in the Filter class. The filter objects form a binary tree, each covering a certain range of the candidate numbers considered. The lack of a math library requires a home-rolled power function (pow() below) and an embedded C-level sqrt() function (Lines 105–106).

The main filter calls the found_prime() function with a prime number. This causes the program to propagate the number found to its children (line 77). This allows them to process the number in parallel with the active object doing a more lengthy operation in cancel_one(), namely iterating over its bit vector and cancelling out all multiples of the found prime.

Once the main active object has found all the primes in $\sqrt{N}$, it calls root_done() which is propagated in a similar fashion as found_prime(). Finally, the done() method is called on each filter active object, which scans the bit vector for any remaining numbers that have not been cancelled out. Those are the prime numbers which are sent to the reporter.

11.3 Example: Preferential Attachment

Preferential attachment is a model of graph construction that produces graphs whose node degrees have a power law distribution. Such graphs model a number of interesting phenomena such as the growth of the World Wide Web or social networks [3,6].

```
18  class Filter
19    bitset:Bitset
20    start:int
21    stop:int
22    first:int
23    report:Reporter
24    adjust:int
25    left:Filter
26    right:Filter
27
28    def pow(n:int, to:int) : int
29      if to == 0 then 1 else n * this.pow(n, to - 1)
30
31    def init(start:int, stop:int, report:Reporter, depth:int) : void {
32      let
33        size = (stop - start) / (this.pow(2, depth) - 1)
34        adjusted_size = if depth > 1 then ((size + 99) / 100 ) * 100 else stop - start
35      in
36        this.bitset = new Bitset(adjusted_size);
37      this.start = start;
38      this.stop = start + this.bitset.size;
39      this.report = report;
40      this.adjust = if start % 2 == 0 then 1 else 0;
41
42      -- print("new Filter({}, {})\n", this.start, this.stop);
43
44      if depth > 1 then
45        let
46          half = (((((stop - this.stop) / 2) + 99) / 100 ) * 100
47        in
48          {
49            this.left = new Filter(this.stop, stop - half, report, depth - 1);
50            this.right = new Filter(stop - half, stop, report, depth - 1);
51          }
52    }
53
54    def calculate_offset(start:int, prime:int) : int
55      let
56        m = prime * prime - start
57        p = if start % prime == 0 then 0 else ((start / prime) * prime + prime) - start
58      in
59        if p < m then m else p
```

Fig. 10. Prime Sieve (b). The `Filter` class (continued in next figure) is the main work horse of this program.

The underlying idea of preferential attachment is that nodes are added incrementally to a graph by establishing connections with existing nodes, such that edges are added from each new node to a random existing node with probability proportional to the degree distribution of the existing nodes. Consequently, the better connected a node is, the higher the chance new nodes will be connected to it (thereby increasing the chance in the future that more new nodes will be connected to it).

The preferential algorithm is based on two parameters: n the total number of nodes in the final graph and k the number of unique edges connecting each

```
61    def cancel_one(prime:int) : void
62      let
63        stop = this.bitset.size
64        i = this.calculate_offset(this.start, prime)
65      in {
66        while i < stop
67          {
68            this.bitset.unset(i);
69            i = i + prime;
70          };
71      }
72
73    def has_children() : bool
74      this.left == null
75
76    def found_prime(p:int) : void {
77      if this.has_children() then { this.left ! found_prime(p); this.right ! found_prime(p); };
78      this.cancel_one(p);
79    }
80
81    def root_done() : void {
82      if this.has_children() then { this.left ! root_done(); this.right ! root_done(); };
83      this.done();
84    }
85
86    def done() : void
87      let
88        i = this.adjust
89        j = this.bitset.size
90        primes = 0
91      in
92        {
93          while i < j
94            {
95              if this.bitset.isset(i) then primes = primes + 1;
96              i = i + 2;
97            };
98          this.report ! report(primes, this.bitset.size);
99        }
```

Fig. 11. Prime Sieve (c). The Filter class (continued from previous figure) is the main work horse of this program.

newly added node to the existing graph. A sequential algorithm for preferential attachment is:

1. Create a fully connected graph of size k (the clique). This is a completely deterministic and all the nodes in the initial graph will be equally likely to be connected to by new nodes.
2. For $i = k+1, \ldots, n$, add a new node n_i, and randomly select k *distinct* nodes from $n_1, \ldots, n_{i-1}$ with probability proportional to the degree of the selected node and add the edges from n_i to the selected nodes to the graph.

One challenge is handling the probabilities correctly. This can be done by storing the edges in an array of size $\approx n \times k$, where every pair of adjacent elements in the array represents an edge. As an example, consider the following graph and its encoding.

```
101  class Main
102    def round_up(i:int) : int
103      ((i + 99) / 100 ) * 100
104
105    def sqrt(i:int) : int
106      embed int sqrt(#{i}); end
107
108    def main() : void
109      let
110        candidates = 1000000000
111        depth = 7
112        stop = this.round_up(this.sqrt(candidates))
113        array = new Bitset(stop)
114        guard = new Reporter(candidates)
115        num = 3
116        idx = 2
117        primes = 1
118        root = new Filter(stop, candidates, guard, depth)
119      in {
120        while num < stop
121          {
122            if array.isset(idx) then {
123              primes = primes + 1;
124              root ! found_prime(num);
125              let
126                j = (num * num) - 1
127              in
128                while j < stop
129                  {
130                    array.unset(j);
131                    j = j + num;
132                  };
133            };
134            num = num + 2;
135            idx = idx + 2;
136          };
137        guard ! report(primes, stop);
138        root ! root_done();
139      }
```

Fig. 12. Prime Sieve (d). The Main class sets up the program and finds all the primes in the first $\sqrt{N}$ (here hard coded to 1 billion) candidates.

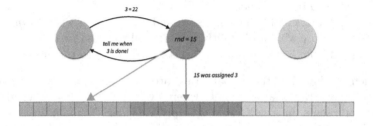

Fig. 13. High-level design of the parallel preferential attachment.

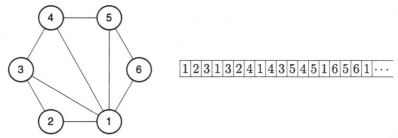

The number of times a node appears in this array divided by the size of the array is precisely the required probability of selecting the node. Thus, when adding new edges, performing a uniform random selection from this array is sufficient to select target nodes.

In the implementation a simple optimisation is made. Half of the edge information is statically known—that is (ignoring the clique), for each $n > k$, the array will look like the following:

$$\cdots \boxed{n} \boxed{m_{n,1}} \boxed{n} \boxed{m_{n,k}} \cdots \boxed{n} \boxed{m_{n,k}} \cdots$$

The indices where the edges for node n will be stored can be calculated in advance, and thus these occurrences of n need not be stored, and the amount of space required can be halved.

Parallelising preferential attachment is non-trivial due to the inherent temporality in the problem: the correctness (with respect to distinctness) of all random choices for an addition depends on the values selected for earlier nodes. However, even though a node may not yet appear in some position in the array, it is possible to compute in parallel a candidate for the desired for *all* future nodes. Then these can gradually be fulfilled (out of order) and the main challenge is ensuring that distinctness of edges is preserved. This is done by checking whenever new edges are added and randomly selecting again when a duplicate edge is added.

The naive implementation shown here attempts to parallelise the algorithm by creating A active objects each responsible for the edges of some nodes in the graph. Every active object proceeds according to the algorithm above, but with non-overlapping start and stop indexes. If a random choice picks an index that is not yet filled in, a message is sent to the active object that owns that part of the array with a request to be notified when that information becomes available.

The requirement that edges are *distinct* needs to be checked whenever a new edge is added. If a duplicate is found, the algorithm just picks another random index. With reasonably large graphs (say 1 million nodes with 20 edges each), duplicate edges is rare, but the scanning is still necessary, and this is more costly in the parallel implementation compared to the sequential one, because in the sequential algorithm all edges are available at the time the test for duplicates is made, but this is not the case in the parallel algorithm.

Figure 13 shows a high-level overview of the implementation. The colour-coded workers own the write rights to the equi-coloured part of the array. A green arrow denotes read rights, a red array denotes write rights. The middle worker attempts to the edge at index 3 in the array, which is not yet determined. This prompts a request to the (left) worker that owns the array chunk to give this information once the requested value becomes available. Once this answer is provided, the middle active object writes this edge into the correct place, provided it is not a duplicate, and forwards the results to any active object that has requested it before this point.

The ENCORE implementation of preferential attachment is fairly long and can be found in Appendix A.

12   Formal Semantics

This section presents the semantics of a fragment of ENCORE via a calculus called μENCORE. The main aim of μENCORE is to formalise the proposed concurrency model of ENCORE, and thereby establish a formal basis for the development of ENCORE and for research on type-based optimisations in the UPSCALE project. μENCORE maintains a strong notion of locality by ensuring that there is no shared data between different active objects in the system.

Syntactic categories.   *Definitions.*

x in Variable

e in Expression

v in Value

$$P ::= \overline{IF}\ \overline{CL}\ e$$

$$T ::= \text{bool} \mid I \mid C \mid \text{void} \mid \text{Fut } T \mid \overline{T} \to T$$

$$IF ::= [\text{passive}]\ \text{interface } I\ \{\ \overline{[Sg]}\ \}$$

$$CL ::= [\text{passive}]\ \text{class } C\ \{\ [\overline{x} : \overline{T};]\ \overline{M}\ \}$$

$$Sg ::= \text{def } m\ ([\overline{x} : \overline{T}]) : T$$

$$M ::= Sg\ \{\ e\ \}$$

$$e ::= e; e \mid x \mid v \mid x.x \mid \text{skip} \mid x = e \mid \text{while } e\ \{\ e\ \} \mid \text{await } e$$
$$\mid\ \text{if } e\ \{\ e\ \}\ \text{else}\ \{\ e\ \} \mid \text{let } \overline{x} = \overline{e}\ \text{in}\ \{\ e\ \} \mid \text{suspend}$$
$$\mid\ e \diamond m(\overline{e}) \mid e \! \downarrow \! m(\overline{e}) \mid \text{new } C\ (\overline{e}) \mid \text{get } e \mid e \rightsquigarrow e$$
$$\mid\ \text{spore } \overline{x} = \overline{e}\ \text{in}\ \lambda(\overline{x} : \overline{T}) \to e : T \mid e(\overline{e})$$

$$v ::= \text{true} \mid \text{false} \mid ()$$

Fig. 14. Syntax of μENCORE. Terms like $\overline{e}$ and $\overline{x}$ denote (possibly empty) lists over the corresponding syntactic categories, square brackets [] denote optional elements.

12.1   The Syntax of μENCORE

The formal syntax of μENCORE is given in Fig. 14. A program P consists of interface and class declarations followed by an expression e which acts as the main block.[8] The types T includes Booleans bool (ignoring other primitive types), a type void (for the () value), type Fut T for futures, interfaces I, passive classes C, and function types $\overline{T} \to T$. In μENCORE, objects can be active or passive. *Active* objects have an independent thread of execution. To store or transfer local data, an active object uses *passive* objects. For this reason, interfaces IF and classes CL can be declared as **passive**. In addition, an interface has a name I and method signatures Sg and class has a name C, fields $\overline{x}$ of type $\overline{T}$, and methods $\overline{M}$. A method signature Sg declares the return type T of a method with

---

[8] μENCORE supports interfaces, though ENCORE does not yet. ENCORE with combine interfaces with traits.

name m and formal parameters $\overline{x}$ of types $\overline{T}$. M defines a method with signature Sg, and expressions e. When constructing a new object of a class C by a statement **new** $C(\overline{e})$, the new object may be either active or passive (depending on the class).

Expressions include, variables (local variables and fields of objects), values, sequential composition $e_1; e_2$, assignment, **skip** (to make semantics easier to write), **if**, **let**, and **while** constructs. Expressions may access the fields of an object, the method's formal parameters and the fields of other objects. *Values* v are expressions on a normal form, *let*-expressions introduce local variables. Cooperative scheduling in μENCORE is achieved by explicitly suspending the execution of the active stack. The statement **suspend** unconditionally suspends the execution of the active method invocation and moves this to the queue. The statement **await** e conditionally suspends the execution; the expression e evaluates to a future f, and execution is suspended only if f has not been fulfilled.

Communication and *synchronisation* are decoupled in μENCORE. In contrast to ENCORE, μENCORE makes explicit in the syntax the two kinds of method call: $e \diamond m(\overline{e})$ corresponds to a synchronous call and $e \wr m(\overline{e})$ corresponds to an asynchronous call. Communication between active objects is based on asynchronous method calls $o \wr m(\overline{e})$ whose return type is Fut T, where T corresponds to the return type of the called method m. Here, o is an object expression, m a method name, and $\overline{e}$ are expressions providing actual parameter values for the method invocation. The result of such a call is a future that will hold the eventual result of the method call. The caller may proceed with its execution *without blocking*. Two operations on futures control synchronisation in μENCORE. The first is **await** f, which was described above. The second is **get** f which retrieves the value stored in the future when it is available, blocking the active object until it is. Futures are first-class citizens of μENCORE. Method calls on passive objects may be synchronous and asynchronous. Synchronous method calls $o \diamond m(\overline{e})$ have a Java-like reentrant semantics. Self calls are written **this** $\wr m(\overline{e})$ or **this** $\diamond m(\overline{e})$.

Anonymous functions and *future chaining* Anonymous functions are available in μENCORE in the form of spores [26]. A spore **spore** $\overline{x}' = \overline{e}'$ **in** $\lambda(\overline{x} : \overline{T}) \rightarrow e : T$ is a form of closure in which the dependencies on local state $\overline{e}'$ are made explicit; i.e., the body e of the lambda-function does not refer directly to variables outside the spore. Spores are evaluated to create closures by binding the variables $\overline{x}'$ to concrete values which are controlled by the closure. This ensures race-free execution even when the body e is not pure. The closure is evaluated by function application $e(\overline{e})$ where the arguments $\overline{e}$ are bound to the variables $\overline{x}$ of the spore, before evaluating the function body e. Closures are first class values. *Future chaining* $e_1 \rightsquigarrow e_2$ allows the execution of a closure e_2 to be spawned into a parallel task, triggered by the fulfilment of a future e_1.

12.2 Typing of μENCORE

Typing judgments are on the form $\Gamma \vdash e : T$, where the typing context Γ maps variables x to their types. (For typing the constructs of the dynamic semantics, Γ will be extended with values and their types.) Write $\Gamma \vdash \overline{e} : \overline{T}$ to denote that

$\Gamma \vdash e_i : T_i$ for $1 \leq i \leq |\overline{e}|$, assuming that $|\overline{e}| = |\overline{T}|$. Types are not assigned to method definitions, class definitions and the program itself; the corresponding type judgements simply express that the constructs are internally well-typed by a tag "**ok**".

Auxiliary definitions. Define function $typeOf(T, m)$ such that: (1) $typeOf(T, m)$ $= \overline{T} \rightarrow T'$ if the method m is defined with signature $\overline{T} \rightarrow T'$ in the class or interface T; (2) $typeOf(T, x) = T'$ if a class T has a field x declared with type T'; and (3) $typeOf(C) = \overline{T} \rightarrow C$ where $\overline{T}$ are the types of the constructor arguments. Further define a predicate $active(T)$ to be true for all active classes and interfaces T and false for all passive classes and interfaces. By extension, let $active(o) = active(C)$ if o is an instance of C.

Subtyping. Let class names C of active classes also be types for the type analysis and let $\preceq$ be the smallest reflexive and transitive relation such that

- $T \preceq$ void for all T,
- $C \preceq I \iff \forall m \in I \cdot typeOf(C, m) \preceq typeOf(I, m)$
- $\overline{T} \preceq \overline{T}' \iff n = length(\overline{T}) = length(\overline{T}')$ and $T_i \preceq T_i'$ for $1 \leq i \leq n$
- $T_1 \rightarrow T_2 \preceq T_1' \rightarrow T_2' \iff T_1' \preceq T_1 \wedge T_2 \preceq T_2'$

A type T is a subtype of T' if $T \preceq T'$. The typing system of μENCORE is given in Fig. 15 and is mostly be standard. Note that rule T-SPORE enforces that all dependencies to the local state to be explicitly declared in the spore.

12.3   Semantics of μENCORE

The semantics of μENCORE is presented as an operational semantics in a context-reduction style [15], using a multi-set of terms to model parallelism (from from rewriting logic [25]).

Run-Time Configurations. The run-time syntax of μENCORE is given in Fig. 16. A *configuration* cn is a multiset of active objects (plus a local heap of passive objects per active object), method invocation messages and futures. The associative and commutative union operator on configurations is denoted by whitespace and the empty configuration by ε.

An active object is given as a term $g(active, hp, q)$ where $active$ is the method invocation that is executing or *idle* if no method invocation is active, hp a multiset of objects, and q is a queue of suspended (or not yet started) method invocations.

An *object obj* is a term $o(\sigma)$ where o is the object's identifier and σ is an assignment of the object's fields to values. Concatenation of such assignments is denoted $\sigma_1 \circ \sigma_2$.

In an *invocation message* $m(o, \overline{v}, hp, f)$, m is the method name, o the callee, $\overline{v}$ the actual parameter values, hp a multiset of objects (representing the data transferred with the message), and f the future that will hold the eventual result. For simplicity, the futures of the system are represented as a mapping *fut* from future identifiers f to either values v or $\perp$ for unfulfilled futures.

$$\frac{}{\Gamma \vdash () : \mathsf{void}} \text{(T-Void)} \qquad \frac{}{\Gamma \vdash \mathbf{true} : \mathsf{bool}} \text{(T-True)} \qquad \frac{}{\Gamma \vdash \mathbf{false} : \mathsf{bool}} \text{(T-False)}$$

$$\text{(T-Await)} \quad \frac{\Gamma \vdash e : \mathsf{Fut}\ T}{\Gamma \vdash \mathbf{await}\ e : \mathsf{void}} \qquad \text{(T-Get)} \quad \frac{\Gamma \vdash e : \mathsf{Fut}\ T}{\Gamma \vdash \mathbf{get}\ e : T} \qquad \text{(T-Var1)} \quad \frac{\Gamma(x) = T}{\Gamma \vdash x : T} \qquad \text{(T-Val)} \quad \frac{\Gamma(v) = T}{\Gamma \vdash v : T} \qquad \text{(T-Sub)} \quad \frac{\Gamma \vdash x : T' \quad T' \preceq T}{\Gamma \vdash x : T}$$

$$\text{(T-Seq)} \quad \frac{\Gamma \vdash e_1 : \mathsf{void} \quad \Gamma \vdash e_2 : T}{\Gamma \vdash e_1; e_2 : T} \qquad \text{(Application)} \quad \frac{\Gamma \vdash \bar{e} : \overline{T} \quad \Gamma \vdash e : \overline{T} \to T}{\Gamma \vdash e(\bar{e}) : T} \qquad \text{(Fut-Chain)} \quad \frac{\Gamma \vdash e' : \mathsf{Fut}\ T' \quad \Gamma \vdash e : T' \to T}{\Gamma \vdash e' \rightsquigarrow e : \mathsf{Fut}\langle T \rangle}$$

$$\text{(T-Assign)} \quad \frac{\Gamma \vdash e : T \quad \Gamma(x) = T}{\Gamma \vdash x = e : T} \qquad \text{(T-Var2)} \quad \frac{\Gamma(x_1) = C \quad typeOf(C, x_2) = T}{\Gamma \vdash x_1.x_2 : T}$$

$$\text{(T-Let)} \quad \frac{\Gamma \vdash \bar{x} = \bar{e}_1 : \overline{T}_1 \quad \Gamma, \bar{x} \mapsto \overline{T}_1 \vdash e_2 : T_2}{\Gamma \vdash \mathbf{let}\ \bar{x} = \bar{e}_1\ \mathbf{in}\ \{e_2\} : T_2} \qquad \text{(T-While)} \quad \frac{\Gamma \vdash e_1 : \mathsf{bool} \quad \Gamma \vdash e_2 : \mathsf{void}}{\Gamma \vdash \mathbf{while}\ e_1\{e_2\} : \mathsf{void}} \qquad \text{(T-Cond)} \quad \frac{\Gamma \vdash e : \mathsf{bool} \quad \Gamma \vdash e_1 : T \quad \Gamma \vdash e_2 : T}{\Gamma \vdash \mathbf{if}\ e\ \{e_1\}\ \mathbf{else}\ \{e_2\} : T}$$

$$\text{(T-AsyncCall)} \quad \frac{typeOf(T', m) = \overline{T} \to T \quad \Gamma \vdash e : T' \quad \Gamma \vdash \bar{e} : \overline{T}}{\Gamma \vdash e\mathord{!} m(\bar{e}) : \mathsf{Fut}\ T}$$

$$\text{(T-SyncCall)} \quad \frac{typeOf(T', m) = \overline{T} \to T \quad \Gamma \vdash e : T' \quad \Gamma \vdash \bar{e} : \overline{T} \quad \neg active(T')}{\Gamma \vdash e \diamond m(\bar{e}) : T}$$

$$\text{(T-New)} \quad \frac{\Gamma \vdash \bar{e} : \overline{T} \quad typeOf(C) = \overline{T} \to C}{\Gamma \vdash \mathbf{new}\ C\ (\bar{e}) : C}$$

$$\text{(T-Spore)} \quad \frac{\Gamma \vdash \bar{e}' : \overline{T}' \quad \bar{x}' : \overline{T}', \bar{x} \mapsto \overline{T} \vdash e : T}{\Gamma \vdash \mathbf{spore}\ \bar{x}' = \bar{e}'\ \mathbf{in}\ \lambda(\bar{x} : \overline{T}) \to e : T\ \overline{T} \to T} \qquad \text{(T-Method)} \quad \frac{\Gamma, \bar{x} \mapsto \overline{T} \vdash e : T}{\Gamma \vdash \mathbf{def}\ m\ (\bar{x} : \overline{T}) : T\ \{e\}\ \mathbf{ok}}$$

$$\text{(T-Class)} \quad \frac{\Gamma' \vdash M\ \mathbf{ok}\ \text{for all}\ M \in \overline{M} \quad \Gamma' = \Gamma, \bar{a}_1 \mapsto \overline{T}_1, \mathbf{this} \mapsto C}{\Gamma \vdash \mathbf{class}\ C\ \{\bar{a}_1 : \overline{T}_1;\ \overline{M}\}\ \mathbf{ok}} \qquad \text{(T-Program)} \quad \frac{\Gamma \vdash e : \mathsf{void} \quad \Gamma \vdash CL\ \text{for all}\ CL \in \overline{CL}}{\Gamma_v \vdash \overline{IF}\ \overline{CL}\ e\ \mathbf{ok}}$$

$$\text{(T-Suspend)} \quad \frac{}{\Gamma \vdash \mathbf{suspend} : \mathsf{void}} \qquad \text{(T-Skip)} \quad \frac{}{\Gamma \vdash \mathbf{skip} : \mathsf{void}}$$

Fig. 15. The type system for μENCORE.

$$
\begin{aligned}
cn &::= \varepsilon \mid msg \mid fut \mid group \mid cn\ cn \\
group &::= g(active, hp, q) \\
hp &::= \varepsilon \mid obj \mid hp \cup hp \\
obj &::= o(\sigma) \\
msg &::= m(o, \bar{v}, hp, f) \\
fut &::= \varepsilon \mid fut[f \mapsto v] \mid fut[f \mapsto \bot] \\
q &::= \emptyset \mid task \mid q \cup q \\
task &::= t(fr \circ sq)
\end{aligned}
\qquad
\begin{aligned}
\sigma &::= \varepsilon \mid x \mapsto \langle T, v \rangle \mid \sigma \circ \sigma \\
active &::= task \mid idle \\
fr &::= \{\sigma | e\} \\
bfr &::= \{\sigma | E\} \\
sq &::= bfr \circ sq \mid \mathsf{eos} \\
v &::= \ldots \mid o \mid f \mid unit \\
&\quad \mid \mathbf{closure}\ \bar{x} = \bar{v}\ \mathbf{in}\ \lambda(\bar{x} : \overline{T}) \to e : T \\
e &::= \ldots \mid e? \\
T &::= \ldots \mid \mathsf{Closure}
\end{aligned}
$$

Fig. 16. Run-time syntax; here, o and f are identifiers for objects and futures, and x is the name of a variable.

The queue q of an active object is a sequence of method invocations (*task*). A *task* is a term $t(fr \circ sq)$ that captures the state of a method invocation as sequence of stack frames $\{\sigma|e\}$ or $\{\sigma|E\}$ (where E is an evaluation context, defined in Fig. 18), each consisting of bindings for local variables plus either the expression being run for the active stack frame or a continuation (represented as evaluation context) for blocked stack frames. *eos* indicates the bottom of the stack. Local variables also include a binding for **this**, the target of the method invocation. The bottommost stack frame also includes a binding for variable *destiny* to the future to which the result of the current call will be stored.

Expressions e are extended with a polling operation $e?$ on futures that evaluates to **true** if the future has been fulfilled or **false** otherwise. Values v are extended with identifiers for the dynamically created objects and futures, and with closures. A *closure* is a dynamically created value obtained by reducing a spore-expression (after sheep cloning the local state). Further assume for simplicity that $default(T)$ denotes a default value of type T; e.g., **null** for interface and class types. Also, classes are not represented explicitly in the semantics, but may be seen as static tables of field types and method definitions. Finally, the run-time type Closure marks the run-time representation of a closure. To avoid introducing too many new run-time constructs, closures are represented as active objects with an empty queue and no fields.

The *initial configuration* of a program reflects its main block. Let o be an object identifier. For a program with main block e the initial configuration consists of a single dummy active object with an empty queue and a task executing the main block itself: $g(t(\{this \mapsto \langle \mathsf{Closure}, o\rangle|e\} \circ eos), o(\varepsilon), \emptyset)$.

A Transition System for Configurations. Transition rules transform configurations cn into new configurations. Let the reflexive and transitive transition relation $\rightarrow$ capture transitions between configurations. A run is a possibly terminating sequence of configurations $cn_0, cn_1, \ldots$ such that $cn_i \rightarrow cn_{i+1}$. Rules apply to subsets of configurations (the standard context rules for configurations are not listed). For simplicity we assume that configurations can be reordered to match the left hand side of the rules, i.e., matching is modulo associativity and commutativity as in rewriting logic [25].

Auxiliary functions. If the class of an object o has a method m, let $bind(m, o, \overline{v})$ and $abind(m, o, \overline{v}, f)$ return a frame resulting from the activation of m on o with actual parameters $\overline{v}$. The difference between these two functions is that the $abind$ introduces a local variable $destiny$ bound to $\langle T, f\rangle$ where T is the return type of the frame. If the binding succeeds, the method's formal parameters are bound to $\overline{v}$. The function $select(q, fut)$ schedules a task which is ready to execute from the task queue q which belongs to an active object g with $g(idle, hp, q)$. The function $atts(C, o)$ returns the initial field assignment σ of a new instance o of class C in which the fields are bound to default values. The function $init(C)$ returns an activation of the *init* method of C, if defined. Otherwise it returns the empty task $\{\varepsilon|()\}$. The predicate $fresh(n)$ asserts that a name n is globally unique (where n may be an identifier for an object or a future). The definition of these functions is straightforward but requires that the class table is explicit in the semantics, which we have omitted for simplicity.

Sheep cloning. Given an object identifier v and a heap hp, let $lookup(v, hp)$ return the object corresponding to v in hp. Given a list of object identifiers $\bar{v}$ and a heap hp, let $rename(\bar{v}, hp, \sigma)$ return a mapping that renames all passive objects reachable from $\bar{v}$ in hp. Given a list of object identifiers $\bar{v}$ and a heap hp, $copy(\bar{v}, hp, transfer)$ returns the sub-heap reachable from $\bar{v}$ in hp. The formal definitions of these functions are given in Fig. 17. Sheep cloning combines the *rename* and *copy* functions.

$$lookup(v, \varepsilon) = \varepsilon$$
$$lookup(v, obj\ hp) = \begin{cases} obj & \text{if } obj = v(a) \\ lookup(v, hp) & \text{otherwise} \end{cases}$$
$$lookup(\textbf{closure } \bar{x}' = \bar{v}' \textbf{ in } \lambda(\bar{x} : \bar{T}) \to e : T, \varepsilon) = \varepsilon$$

$$rename(\varepsilon, hp, \sigma) = \sigma$$
$$rename(v\,\bar{v}, hp, \sigma) = rename(\bar{v}, hp, \sigma) \text{ if } \begin{cases} v \in dom(\sigma), \\ lookup(v, hp) = v(a) \wedge active(v), \text{or} \\ v \text{ is neither an object nor a closure} \end{cases}$$
$$rename(v\,\bar{v}, hp, \sigma) = rename(ran(a)\,\bar{v}, hp, \sigma[v \mapsto v'])$$
$$\qquad \text{if } lookup(v, hp) = v(a) \wedge \neg active(v) \wedge fresh(v') \wedge \neg active(v')$$
$$rename((\textbf{closure } \bar{x}' = \bar{v}' \textbf{ in } \lambda(\bar{x} : \bar{T}) \to e : T)\,\bar{v}, hp, \sigma) = rename(\bar{v}'\,\bar{v}, hp, \sigma)$$

$$copy(\varepsilon, hp, transfer) = transfer$$
$$copy(v\,\bar{v}, hp, transfer) = transfer) \text{ if } \begin{cases} lookup(v, hp) \in transfer, \\ lookup(v, hp) = v(a) \wedge active(v), \text{or} \\ v \text{ is neither an object nor a closure} \end{cases}$$
$$copy(v\,\bar{v}, hp, transfer) = copy(ran(a)\,\bar{v}, hp, transfer \cup \{v(a)\})$$
$$\qquad \text{if } lookup(v, hp) = v(a) \wedge \neg active(v)$$
$$copy((\textbf{closure } \bar{x}' = \bar{v}' \textbf{ in } \lambda(\bar{x} : \bar{T}) \to e : T)\,\bar{v}, hp, transfer) = copy(\bar{v}'\,\bar{v}, hp, transfer)$$

Fig. 17. Sheep cloning: deep renaming and copying of passive objects.

Transition rules. The transition rules of μENCORE are presented in Figs. 20 and 21. Let a denote the map of fields to values in an object and l to denote map of local variables to values in a (possibly blocked) frame. A context reduction semantics decomposes an expression into a reduction context and a redex, and reduces the redex (e.g., [15,28]). A *reduction context* is denoted by an expression E with a single hole denoted by $\bullet$, while an expression without any holes is denoted by e. Filling the hole of a context E with an expression e is denoted by $E[e]$, which represents the expression obtained by replacing the hole of E with e. In the rules, an expression $E[e]$ consisting of a context E and a redex e is reduced to $E[e']$, possibly with side effects. Here the context E determines the hole where a reduction may occur and e is the redex located in the position of

$$E ::= \bullet \mid E; e \mid E? \mid \text{if } E \{ e_1 \} \text{else} \{ e_2 \} \mid x = E \mid \text{let } \overline{x} = \overline{v}, E, \overline{e} \text{ in } \{ e \}$$
$$\mid E \diamond m(\overline{e}) \mid v \diamond m(\overline{v}, E, \overline{e}) \mid E \sharp m(\overline{e}) \mid v \sharp m(\overline{v}, E, \overline{e}) \mid \text{new } C (\overline{v}, E, \overline{e}) \mid \text{get } E$$
$$\mid \text{spore } \overline{x} = \overline{v}, E, \overline{e} \text{ in } \lambda(\overline{x} : \overline{T}) \rightarrow e : T \mid E(\overline{e}) \mid v(\overline{v}, E, \overline{e}) \mid E \rightsquigarrow e \mid v \rightsquigarrow E$$

Fig. 18. Context reduction semantics of μENCORE: the contexts.

$$redexes ::= x \mid x.x \mid v \mid v; e \mid v? \mid \text{if } v \{ e_1 \} \text{else} \{ e_2 \} \mid \text{await } e \mid x = v \mid \text{while } e_1 \{ e_2 \}$$
$$\mid \text{let } \overline{x} = \overline{v} \text{ in } \{ e \} \mid v \diamond m(\overline{v}) \mid v \sharp m(\overline{v}) \mid \text{new } C (\overline{v}) \mid \text{get } v \mid \text{skip} \mid v \rightsquigarrow v'$$
$$\mid \text{closure } \overline{x}' = \overline{v}' \text{ in } (\lambda(\overline{x} : \overline{T}) \rightarrow e : T)(\overline{v}) \mid \text{spore } \overline{x}' = \overline{v}' \text{ in } (\lambda(\overline{x} : \overline{T}) \rightarrow e : T)$$

Fig. 19. Context reduction semantics of μENCORE: the redexes.

that hole. The contexts of our semantics are given in Fig. 18 and the redexes in Fig. 19.

Basic rules. Rule SKIP consumes a **skip** in the active task. Rules ASSIGN1 and ASSIGN2 assign a value v to a variable x in the local variables l or in the fields a, respectively. In the rules, the premise $l(this) = o$ looks up the corresponding object. Rule VARIABLE1 reads the value of a local variable or a field of the object executing the frame. Rule VARIABLE2 reads the value of the field of another object in the same local heap. Rules COND1 and COND2 cover the two cases of conditional expression. Rule LET associates a value v to a local variables x and uses it in the expression e. Rule WHILE unfolds the **while** loop into a conditional. Rule SEQUENTIAL discards the value v in an expression of the form $v; e$ and continues the evaluation the expression e.

Suspension and activation. Rule SUSPEND enables cooperative scheduling and moves the active task to the queue q, making the active task *idle*. Rule AWAIT unfolds into a conditional. Rules POLL-FUTURE1 and POLL-FUTURE2 test whether the future f has been resolved. If f is not resolved, the active task suspends. Otherwise, the **await** expression is reduced to a **skip**. Rule ACTIVATE schedules a task from the task queue q by means of the *select* function. Since the schedulability of a task may depend on a future, *select* uses the map of futures *fut*.

Asynchronous method calls. Rule REMOTE-ASYNC-CALL sends an invocation message to an object o, with a return address f and sheep copied actual parameter values. The cloned objects hp' are transferred with the method invocation. (Notation $\overline{v} \sigma$ and $hp \sigma$ denote the recursive application of the substitution σ to $\overline{v}$ and hp, respectively). The identifier of the new future is added to *fut* with a $\bot$ value indicating that it has not been fulfilled. In rule REMOTE-BIND-MTD, the function $abind(m, o, \overline{v}, f)$ binds a method call in the class of the callee o. This results in a new task in the queue of the active object o. In the frame fr, the local variable *this* is bound to o and *destiny* is bound to f. The heap hp' transferred with the message extends the heap hp of the active object. Rule LOCAL-ASYNC-CALL puts a new task with a single frame fr on the queue q. As before, a new future f is created and associated to the variable *destiny* in fr, the identifier of the new future is added to *fut* with a $\bot$ value. Rule ASYNC-RETURN places the final value

$$
\begin{array}{ccc}
(\textsc{Skip}) & (\textsc{Suspend}) & (\textsc{Sequential}) \\
g(t(\{l|E[\textbf{skip}]\} \circ sq), hp, q) & g(t(\{l|E[\textbf{suspend}]\} \circ sq), hp, q) & g(t(\{l|E[v; e]\} \circ sq), hp, q) \\
\to g(t(\{l|E[()]\} \circ sq), hp, q) & \to g(idle, hp, t(\{l|E[()]\} \circ sq) \circ q) & \to g(t(\{l|E[e]\} \circ sq), hp, q)
\end{array}
$$

$$
(\textsc{Variable1}) \qquad\qquad\qquad (\textsc{Let})
$$
$$
\frac{l(this) = o \quad o(a) \in hp \quad a \circ l(x) = \langle T, v \rangle}{\begin{array}{c} g(t(\{l|E[x]\} \circ sq), hp, q) \\ \to g(t(\{l|E[v]\} \circ sq), hp, q) \end{array}} \qquad \frac{fresh(\overline{y})}{\begin{array}{c} g(t(\{l|E[\textbf{let } \overline{x} = \overline{v} \textbf{ in } \{ e \}]\} \circ sq), hp, q) \\ \to g(t(\{l[\overline{y} \mapsto \overline{v}]|E[e[\overline{x}/\overline{y}]]\} \circ sq), hp, q) \end{array}}
$$

$$
(\textsc{Variable2}) \qquad\qquad\qquad (\textsc{Activate})
$$
$$
\frac{\{o(a), o'(a')\} \subseteq hp \quad l(this) = o \quad a \circ l(x_1) = \langle T', o' \rangle \quad a' \circ l(x_2) = \langle T, v \rangle}{\begin{array}{c} g(t(\{l|E[x_1.x_2]\} \circ sq), hp, q) \\ \to g(t(\{l|E[v]\} \circ sq), hp, q) \end{array}} \qquad \frac{\begin{array}{c} task = select(q, fut) \\ q' = q \setminus \{task\} \end{array}}{\begin{array}{c} g(idle, hp, q) \; fut \\ \to g(task, hp, q') \; fut \end{array}}
$$

$$
(\textsc{Poll-Future1}) \qquad\qquad\qquad (\textsc{Poll-Future2})
$$
$$
\frac{fut(f) = v}{\begin{array}{c} g(t(\{l|E[f?]\} \circ sq), hp, q) \; fut \\ \to g(t(\{l|E[\textbf{true}]\} \circ sq), hp, q) \; fut \end{array}} \qquad \frac{fut(f) = \bot}{\begin{array}{c} g(t(\{l|E[f?]\} \circ sq), hp, q) \; fut \\ \to g(t(\{l|E[\textbf{false}]\} \circ sq), hp, q) \; fut \end{array}}
$$

$$
(\textsc{New-Active-Object}) \qquad\qquad\qquad (\textsc{New-Passive-Object})
$$
$$
\frac{\begin{array}{c} a = atts(C, o) \quad fresh(o) \quad fresh(f) \\ active(C) \quad fut' = fut[f \mapsto \bot] \quad \{l'|e'\} = init(C, f) \end{array}}{\begin{array}{c} g(t(\{l|E[\textbf{new } C()]\} \circ sq), hp, q) \; fut \\ \to g(t(\{l|E[o]\} \circ sq), hp, q) \; fut' \\ g(t(\{l'[this \mapsto \langle C, o \rangle]|e'\} \circ eos), \{o(a)\}, \emptyset) \end{array}} \qquad \frac{\begin{array}{c} fr = \{l'[this \mapsto \langle C, o \rangle]|e'; o\} \\ \{l'|e'\} = init(C) \quad hp' = hp \cup \{o(a)\} \\ \neg active(C) \quad fresh(o) \quad a = atts(C, o) \end{array}}{\begin{array}{c} g(t(\{l|E[\textbf{new } C()]\} \circ sq), hp, q) \\ \to g(t(fr \circ \{l|E[\bullet]\} \circ sq), hp', q) \end{array}}
$$

$$
(\textsc{Assign1}) \qquad\qquad\qquad (\textsc{Assign2})
$$
$$
\frac{l(x) = \langle T, v' \rangle \quad l' = l[x \mapsto \langle T, v \rangle]}{\begin{array}{c} g(t(\{l|E[x = v]\} \circ sq), hp, q) \\ \to g(t(\{l'|E[()]\} \circ sq), hp, q) \end{array}} \qquad \frac{\begin{array}{c} a(x) = \langle T, v' \rangle \quad x \notin dom(l) \\ hp' = hp \cup \{o(a[x \mapsto \langle T, v \rangle])\} \quad l(this) = o \end{array}}{\begin{array}{c} g(t(\{l|E[x = v]\} \circ sq), hp \cup \{o(a)\}, q) \\ \to g(t(\{l|E[()]\} \circ sq), hp', q) \end{array}}
$$

$$
(\textsc{Cond1}) \qquad\qquad\qquad (\textsc{Cond2})
$$
$$
\begin{array}{cc}
g(t(\{l|E[\textbf{if true } \{e_1\} \textbf{ else } \{e_2\}]\} \circ sq), hp, q) & g(t(\{l|E[\textbf{if false } \{e_1\} \textbf{ else } \{e_2\}]\} \circ sq), hp, q) \\
\to g(t(\{l|E[e_1]\} \circ sq), hp, q) & \to g(t(\{l|E[e_2]\} \circ sq), hp, q)
\end{array}
$$

$$
(\textsc{Await}) \qquad\qquad\qquad (\textsc{While})
$$
$$
\begin{array}{cc}
g(t(\{l|E[\textbf{await } e]\} \circ sq), hp, q) & g(t(\{l|E[\textbf{while } e_1 \{ e_2 \}]\} \circ sq), hp, q) \\
\to g(t(\{l|E[\textbf{if } e? \{\textbf{skip}\} \textbf{ else} & \to g(t(\{l|E[\textbf{if } e_1 \{e_2; \textbf{while } e_1 \{ e_2 \}\} \\
\{\textbf{suspend; await } e\}]\} \circ sq) \,, hp, q) & \textbf{else } \{\textbf{skip}\}]\} \circ sq), hp, q)
\end{array}
$$

Fig. 20. Semantics for μENCORE (1).

of a task into the associated future, making the active task idle. Rule READ-FUT dereferences a future f from the maps of futures fut.

Synchronous method calls. In rule SYNC-CALL, method m is called on a local object o, with actual parameters $\overline{v}$. The function $bind(m, o', \overline{v}')$ binds the call in the class of o, resulting in a frame fr. The new frame extends the stack, and the previously active frame becomes blocked. In rule SYNC-RETURN, the active frame

$$\text{(Local-Async-Call)}$$
$$\frac{fut' = fut[f \mapsto \bot] \quad abind(m, o, \bar{v}, f) = fr}{o \in dom(hp) \quad fresh(f) \quad q' = t(fr \circ \text{eos}) \circ q} $$
$$g(t(\{l|E[o \mathbin{\text{\textslash}} m(\bar{v})]\} \circ sq), hp, q) \, fut$$
$$\rightarrow g(t(\{l|E[f]\} \circ sq), hp, q') \, fut'$$

$$\text{(Remote-Async-Call)}$$
$$\frac{\begin{array}{c} o \notin dom(hp) \quad fresh(f) \\ \bar{v}' = \bar{v}\sigma \quad rename(\bar{v}, hp, \varepsilon) = \sigma \\ fut' = fut[f \mapsto \bot] \quad copy(\bar{v}', hp\,\sigma, \varepsilon) = hp' \end{array}}{\begin{array}{c} g(t(\{l|E[o \mathbin{\text{\textslash}} m(\bar{v})]\} \circ sq), hp, q) \, fut \\ \rightarrow g(t(\{l|E[f]\} \circ sq), hp, q) \\ m(o, \bar{v}', hp', f) \, fut' \end{array}}$$

$$\text{(Sync-Call)}$$
$$\frac{o \in dom(hp) \quad bind(m, o, \bar{v}) = fr}{g(t(\{l|E[o \diamond m(\bar{v})]\} \circ sq), hp, q)}$$
$$\rightarrow g(t(fr \circ \{l|E[\bullet]\} \circ sq), hp, q)$$

$$\text{(Remote-Bind-Mtd)}$$
$$\frac{o(a) \in hp \quad q' = \{t(fr \circ \text{eos})\} \cup q}{fresh(s) \quad abind(m, o, \bar{v}, f) = fr}$$
$$g(active, hp, q) \, m(o, \bar{v}, hp', f)$$
$$\rightarrow g(active, hp \cup hp', q')$$

$$\text{(Async-Return)}$$
$$\frac{f = l(destiny) \quad fut' = fut[f \mapsto v]}{g(t(\{l|v\} \circ \text{eos}), hp, q) \, fut}$$
$$\rightarrow g(idle, hp, q) \, fut'$$

$$\text{(Read-Fut)}$$
$$\frac{fut(f) = v \quad v \neq \bot}{g(t(\{l|E[\text{get } f]\} \circ sq), hp, q) \, fut}$$
$$\rightarrow g(t(\{l|E[v]\} \circ sq), hp, q) \, fut$$

$$\text{(Sync-Return)}$$
$$g(t(\{l|v\} \circ \{l'|E[\bullet]\} \circ sq), hp, q)$$
$$\rightarrow g(t(\{l'|E[v]\} \circ sq), hp, q)$$

$$\text{(Create-Closure)}$$
$$\frac{\bar{v}' = \bar{v}\sigma \quad rename(\bar{v}, hp, \varepsilon) = \sigma \quad copy(\bar{v}', hp\,\sigma, \varepsilon) = hp'}{g(t(\{l \mid E[(\text{spore } \bar{x}' = \bar{v} \text{ in } \lambda(\bar{x} : \overline{T}) \rightarrow e : T)]\} \circ sq), hp, q)}$$
$$\rightarrow g(t(\{l \mid E[(\text{closure } \bar{x}' = \bar{v}' \text{ in } \lambda(\bar{x} : \overline{T}) \rightarrow e : T)]\} \circ sq), hp \cup hp', q)$$

$$\text{(Lambda-App)}$$
$$g(t(\{l \mid E[(\text{closure } \bar{x}' = \bar{v}' \text{ in } \lambda(\bar{x} : \overline{T}) \rightarrow e : T)(\bar{v})]\} \circ sq), hp, q)$$
$$\rightarrow g(t(\{l \mid E[\text{let } \bar{x}', \bar{x} = \bar{v}', \bar{v} \text{ in } \{ e \}]\} \circ sq), hp, q)$$

$$\text{(New-Chained-Object)}$$
$$\frac{\begin{array}{c} \bar{v}' = \bar{v}\sigma \quad hp' = copy(\bar{v}', hp\,\sigma, \varepsilon) \cup \{o(\varepsilon)\} \\ rename(\bar{v}, hp, \varepsilon) = \sigma \quad v_1 = \text{closure } \bar{x} = \bar{v} \text{ in } \lambda(x : T) \rightarrow e : T' \\ fresh(o) \quad fresh(f') \quad v_2 = \text{closure } \bar{x} = \bar{v}' \text{ in } \lambda(x : T) \rightarrow e : T' \\ l' = [this \mapsto \langle \text{Closure}, o \rangle, destiny \mapsto \langle \text{Fut } T', f' \rangle] \quad fut' = fut[f' \mapsto \bot] \end{array}}{g(t(\{l|E[f \rightsquigarrow v_1]\} \circ sq), hp, q) \, fut}$$
$$\rightarrow g(t(\{l|E[f']\} \circ sq), hp, q) \, g(t(\{l'|v_2 \, (\text{get } f)\} \circ \text{eos}), \{hp'\}, \emptyset) \, fut'$$

Fig. 21. Semantics for μEncore (2).

only contains a single value. The frame is popped form the stack, and the value is passed to the blocked frame below which becomes active.

Object creation. Rule New-Active-Object creates a an active object with a unique identifier o' and a new local heap. The fields of o' are given default values by $atts(C, o')$. The active task of the new active object is the constructor $init(C, f)$, where the local variable *this* binds to $\langle C, o \rangle$ and *destiny* binds to $\langle \text{Fut } C, f \rangle$.[9] Passive object are created in a similar way, except that the class constructor is executed by the calling thread (cf. Rule New-Passive-Object).

---

[9] In Encore the constructor does not run asynchronously.

Closures. In rule CREATE-CLOSURE, a closure is created from a spore by sheep cloning any references to the passive objects of enclosing active object. Note that values inside closures are also sheep copied, even if this has already been done when they were created, to ensure that if the closure is passed out of the active object, it is passed with a fresh sheep clone of its passive objects. Rule LAMBDA-APP reduces a closure to a let-expression when it is applied to values.

Future chaining. Future chaining creates a new dummy active object in which the closure can execute in parallel with the current active object. The closure blocks waiting for the value of it needs from f to begin execution, and will return its own value to another future f'.

12.4 Run-Time Typing

Assume a typing context $CT(C)$ that maps fields of each class C to their declared types. The class table also includes a class Closure such that $CT(\text{Closure}) = \varepsilon$. The run-time type system (Fig. 22) facilitates the type preservation proof for μENCORE.

$$(\text{RT-EMPTY}) \qquad (\text{RT-IDLE}) \qquad (\text{RT-OID}) \qquad (\text{RT-QUEUE1})$$
$$\Gamma \vdash \varepsilon \text{ ok} \qquad \Gamma \vdash \text{idle ok} \qquad \Gamma \vdash v : \Gamma(v) \qquad \Gamma \vdash \emptyset \text{ ok}$$

$$(\text{RT-POLL}) \qquad (\text{RT-FUT1}) \qquad \frac{(\text{RT-FUT2})}{\Gamma \vdash v : T}$$
$$\frac{\Gamma \vdash e : \text{Fut}\langle T\rangle}{\Gamma \vdash e? : \text{Bool}} \qquad \frac{\Gamma \vdash \textit{fut} \text{ ok}}{\Gamma \vdash \textit{fut}[f \mapsto \bot] \text{ ok}} \qquad \frac{\Gamma \vdash \textit{fut} \text{ ok} \quad \Gamma(f) = \text{Fut } T}{\Gamma \vdash \textit{fut}[f \mapsto v] \text{ ok}} \qquad \frac{(\text{RT-TASK})}{\Gamma \vdash fr \circ sq : \text{Unit}}{\Gamma \vdash s(fr \circ sq) \text{ ok}}$$

$$(\text{RT-STACK2}) \qquad (\text{RT-QUEUE2}) \qquad (\text{RT-SUBST1}) \qquad (\text{RT-SUBST2})$$
$$\frac{\Gamma \vdash \{l|e\} : T \quad l(destiny) = \langle \text{Fut}\langle T\rangle, f\rangle}{\Gamma \vdash \{l|e\} \circ eos : T} \qquad \frac{\Gamma \vdash q_1 \text{ ok} \quad \Gamma \vdash q_2 \text{ ok}}{\Gamma \vdash q_1 \cup q_2 \text{ ok}} \qquad \frac{\Gamma(x) = T \quad \Gamma \vdash v : T}{\Gamma \vdash x \mapsto \langle T, v\rangle \text{ ok}} \qquad \frac{\Gamma \vdash \sigma_1 \text{ ok} \quad \Gamma \vdash \sigma_2 \text{ ok}}{\Gamma \vdash \sigma_1 \circ \sigma_2 \text{ ok}}$$

$$(\text{RT-FRAME}) \qquad\qquad (\text{RT-MSG})$$
$$\frac{\sigma(this) = \langle C, o\rangle \quad \Gamma' \vdash \sigma \text{ ok} \quad \Gamma' \vdash e : T \quad \sigma = \overline{x} \mapsto \langle \overline{T}, \overline{v}\rangle \quad \Gamma' = \Gamma \circ CT(C) \circ \overline{x} \mapsto \overline{T}}{\Gamma \vdash \{\sigma|e\} : T} \qquad \frac{\Gamma \vdash o : T' \quad \Gamma \vdash \overline{v} : \overline{T} \quad typeOf(T', m) = \overline{T} \to T \quad \Gamma \vdash f : \text{Fut } T \quad \Gamma \vdash hp \text{ ok}}{\Gamma \vdash m(o, \overline{v}, hp, f) \text{ ok}}$$

$$(\text{RT-OBJ}) \qquad (\text{RT-HEAP}) \qquad (\text{RT-CONF}) \qquad (\text{RT-GROUP})$$
$$\frac{\Gamma(o) = C \quad \Gamma \circ CT(C) \vdash \sigma \text{ ok}}{\Gamma \vdash o(\sigma) \text{ ok}} \qquad \frac{\Gamma \vdash hp_1 \text{ ok} \quad \Gamma \vdash hp_2 \text{ ok}}{\Gamma \vdash hp_1 \cup hp_2 \text{ ok}} \qquad \frac{\Gamma \vdash cn_1 \text{ ok} \quad \Gamma \vdash cn_2 \text{ ok}}{\Gamma \vdash cn_1 \ cn_2 \text{ ok}} \qquad \frac{\Gamma \vdash active : T \quad \Gamma \vdash hp \text{ ok} \quad \Gamma \vdash q \text{ ok}}{\Gamma \vdash g(active, hp, q) \text{ ok}}$$

$$(\text{RT-CLOSURE})$$
$$\frac{\Gamma \vdash (\textbf{spore } \overline{x} = \overline{v} \textbf{ in } \lambda(\overline{x}' : \overline{T}) \to e : T) : \overline{T} \Rightarrow T}{\Gamma \vdash (\textbf{closure } \overline{x} = \overline{v} \textbf{ in } \lambda(\overline{x}' : \overline{T}) \to e : T) : \overline{T} \Rightarrow T}$$

Fig. 22. Type system for μENCORE run-time states.

Lemma 1. *If a program $\overline{IF}\ \overline{CL}\ e$ is well-typed, then there is a Γ such that the initial run-time state of this program is well-typed:* $\Gamma \vdash s(\{\textit{this} \mapsto \langle \mathsf{Closure}, o \rangle | e \} \circ eos, o(\varepsilon), \emptyset)$ **ok***.*

Lemma 2 (Sheep lemma). *Assume that $\Gamma \vdash hp$ **ok** and let σ be a substitution such that $dom(\sigma) \subseteq dom(\Gamma)$, $ran(\sigma) \cap dom(\Gamma) = \emptyset$. Let $\Gamma' = \{y \mapsto T | \sigma(x) = y \wedge \Gamma(x) = T\}$ Then $\Gamma \circ \Gamma' \vdash hp\,\sigma$* **ok***.*

Lemma 3 (Type preservation). *If $\Gamma \vdash cn$ **ok** and $cn \to cn'$ then there exists a Γ' such that $\Gamma \subseteq \Gamma'$ and $\Gamma \vdash cn'$* **ok**

Theorem 1. *Let P be a program in μENCORE, with an initial state cn. If $\Gamma \vdash P$* **ok** *and $cn \to cn'$, there is a typing environment Γ' such that $\Gamma \subseteq \Gamma'$ and $\Gamma' \vdash cn'$* **ok***.*

Proof. Follows directly from Lemmas 1–3.

13   Related Work

ENCORE is based on the concurrency model provided by active objects and actor-based computation, where software units with encapsulated thread of control communicate asynchronously. Languages based on the actor model [1,2] take asynchronous messages as the communication primitive and focus on loosely coupled processes with less synchronisation. This makes actor languages conceptually attractive for parallel programming programming. Rather than the pure asynchronous message passing model of actor systems, active objects adopts method calls as asynchronous messages combined with futures to deliver results. Futures were devised as a means to reduce latency by synchronising at the latest possible time. Futures were discovered by Baker and Hewitt in the 70s [5], and rediscovered after around 10 years and introduced in languages such as ABCL [40,41], Argus [24], ConcurrentSmalltalk [39], and MultiLisp [18] and later in Alice [33], Oz-Mozart [36], Concurrent ML [32], C++ [23] and Java [37], often as libraries. Nowadays, active object and actor-based concurrency is increasingly attracting attention due to its intuitive and compositional nature, which can lead to good scalability in a parallel setting. Modern example languages or frameworks include Erlang [4], ProActive [7], Scala Actors [17], Kilim [34], Creol [21], ABS [20], Akka [35], Habanero-Java [9], among others.

ENCORE has a clear distinction between active and passive objects, such that passive objects as a default are only locally accessible. This is ensured in μENCORE by means of sheep cloning [11] and paves the way for capability type systems for sharing, as first investigated in Joëlle [13]. ENCORE further features spores, originally proposed for Scala [26]. Although spores in ENCORE need not be pure, they are combined with sheep cloning to preserve race-free execution as a default. Future versions of spores in ENCORE will utilise capabilities for more fine-grained control.

14   Conclusion

Programming parallel computers is hard, but as all future computers will be parallel computers doing so will be a necessary skill of all programmers. New programming models and languages are required to support programmers in writing applications that are safe and exploit the available parallel computing resources. ENCORE aims to answer this challenge by provided active-object based parallelism combined with additional mechanisms such as parallel combinators and capabilities for safely expressing other forms of parallelism. This paper gave an overview of ENCORE, including the semantics of its core, along with a number of examples showing how to use the language.

Work on ENCORE has really only just begun. In the future we will be implementing and improving language constructs for expressing different kinds of parallelism and for controlling sharing, data layout and other deployment related concerns. We will continue improving the run-time system, developing libraries and tool support, and exploring case studies. In the near future, we plan to convert the compiler to open source. When this happens—or even beforehand, if you are keen—, you are more than welcome to contribute to the development of ENCORE.

A   Code for Preferential Attachments

```
1  class Worker
2    id : int
3    k : int
4    owners_map : Map<int,Worker>
5    n_size : int
6    n_start : int
7    outstanding : int
8    edges : Set<int>
9    home : Main
10   connections : Array<int>
11
12   def init(id:int, k:int, cs:Array<int>, map:Map<int,Worker>, m:Main) : void {
13     this.id = id;
14     this.k = k;
15     this.owners_map = map;
16     this.n_size = map.range / k;
17     this.n_start = this.n_size * id;
18     this.edges = new Set<int>(k);
19     this.home = m;
20     this.connections = cs;
21   }
22
23   def check_done() : void
24     if this.outstanding == 0 then this.home ! done(this.id)
25
26   def start() : void {
```

```
27    let
28      start = this.n_start
29    in
30      repeat i <- this.n_size
31        this.add_node(start + i);
32
33    this.check_done()
34  }
35
36  def add_node(n:int) : void {
37    this.edges.reset();
38
39    let
40      start = n * this.k
41    in
42      repeat i <- this.k
43        this.add_edge(n, start + i)
44  }
45
46  def add_edge(n:int, target_i:int) : void
47    let
48      coin_flip = this.random(0, 2)
49      from_i = this.random(0, n * this.k)
50      c = if coin_flip == 0 then (from_i % n) + 1
51                            else this.connections.read(from_i)
52    in
53          if c < 0 then this.add_edge(n, target_i)
54      else if c == 0 then this.add_remote_edge(target_i, from_i);
55      else this.add(n, target_i, c);
56
57  def add_remote_edge(local_i:int, remote_i:int) : void {
58    this.outstanding = this.outstanding + 1;
59    this.remote_actor(remote_i) ! read_location(remote_i, local_i, this);
60  }
61
62  def read_location(local_i:int, remote_i:int, from:Worker) : void
63    let
64      c = this.connections.read(local_i)
65    in
66          if c > 0 then from ! add_edge_from_remote(remote_i, c, this)
67      else if c == 0 then this ! read_location(local_i, remote_i, from)
68
69  def redo(i:int) : void {
70    this.add_edge(i / this.k, i);
71    this.outstanding = this.outstanding - 1;
72  }
73
74  def add_edge_from_remote(i:int, c:int, from:Worker) : void {
75    this.outstanding = this.outstanding - 1;
76    this.edges.reset();
77
78    let start_i = i - (i % this.k) in repeat j <- this.k
79      let
80        v = this.connections.read(start_i + j)
```

```
81        in
82          if not this.edges.add(v) then unless v == 0 then
83            this.add_edge(i / this.k, start_i + j);
84
85        if this.edges.add(c) then this.add_local_edge(i, c)
86                          else this.add_edge(i / this.k, i);
87
88        this.check_done();
89      }
90
91      def add_local_edge(i:int, c:int) : void
92        this.connections.write(i, c)
93
94      def remote_actor(i:int) : Worker
95        this.owners_map.lookup(i / this.k)
96
97      def add(n:int, i:int, c:int) : void
98          if c < 0 then this.add_local_edge(i, c)
99        else if this.edges.add(c) then this.add_local_edge(i, c)
100       else this.add_edge(n, i)
101
102     def random(a:int,b:int) : int
103       embed int
104         (random() % b) + a;
105       end
106
107     def generate_clique() : void {
108       repeat n <- this.k
109         repeat k <- this.k
110           let i = k + (n * this.k) in
111             this.add_local_edge(i, if k < n then k+1 else (0-1));
112
113       repeat n <- (this.n_size - this.k)
114         this.add_node(this.k + n);
115
116       this.home ! done(this.id);
117     }
118
119 class Main
120   workers:int
121
122     def done(id:int) : void {
123       this.workers = this.workers - 1;
124       if this.workers == 0 then print("Done!\n");
125     }
126
127     def fix_mod_assert(n:int, k:int, w:int) : int
128       (n * k) % w
129
130     def main() : void
131       let
132         n = 1000000
133         k = 32
```

```
134    workers = 32
135    array = new Array<int>(n * k);
136    owners_map = new Map<int,Worker>(n * k, workers);
137  in
138    {
139      embed void
140        if (argc > 3) {
141          #{n} = atoi(argv[1]);
142          #{k} = atoi(argv[2]);
143          #{workers} = atoi(argv[3]);
144        }
145      end;
146      -- To avoid stupid rounding errors when splitting
147      -- the array across actors
148      assertTrue(this.fix_mod_assert(n, k, workers) == 0);
149      assertTrue(n > k * workers);
150      repeat i <- workers
151        let
152          a = new Worker(i, k, array, owners_map, this)
153        in
154          {
155            this.owners_map.add(a);
156
157            if i == 0 then a ! generate_clique()
158                      else a ! start();
159          }
160    }
```

References

1. Agha, G.A.: ACTORS: A Model of Concurrent Computations in Distributed Systems. The MIT Press, Cambridge (1986)
2. Agha, G.A., Mason, I.A., Smith, S.F., Talcott, C.L.: A foundation for actor computation. J. Funct. Program. **7**(1), 1–72 (1997)
3. Albert, R., Barabási, A.-L.: Statistical mechanics of complex networks. Rev. Mod. Phys. **74**, 47–97 (2002)
4. Armstrong, J.: Programming Erlang: Software for a Concurrent World. Pragmatic Bookshelf, Raleigh (2007)
5. Baker Jr, H.C., Hewitt, C.: The incremental garbage collection of processes. SIGPLAN Not. **12**(8), 55–59 (1977)
6. Barabási, A.-L., Albert, R.: Emergence of scaling in random networks. Science **286**(5439), 509–512 (1999)
7. Caromel, D., Henrio, L., Serpette, B.P.: Asynchronous sequential processes. Inf. Comput. **207**(4), 459–495 (2009)
8. Castegren, E., Wrigstad, T.: Capable: capabilities for scalability. In: IWACO 2014 (2014)
9. Cavé, V., Zhao, J., Shirako, J., Sarkar, V.: Habanero-Java: the new adventures of old X10. In: Probst, C.W., Wimmer, C. (eds.) Proceedings of the 9th International Conference on Principles and Practice of Programming in Java, PPPJ 2011, pp. 51–61. ACM, Kongens Lyngby, Denmark, 24–26 August 2011
10. Clarke, D.: Object ownership and containment. Ph.D. thesis, School of Computer Science and Engineering, University of New South Wales, Australia (2002)

11. Clarke, D., Noble, J., Wrigstad, T. (eds.): Aliasing in Object-Oriented Programming. Types, Analysis and Verification. LNCS, vol. 7850. Springer, Heidelberg (2013)
12. Clarke, D., Wrigstad, T.: External uniqueness is unique enough. In: Cardelli, L. (ed.) ECOOP 2003. LNCS, vol. 2743, pp. 176–200. Springer, Heidelberg (2003)
13. Clarke, D., Wrigstad, T., Östlund, J., Johnsen, E.B.: Minimal ownership for active objects. In: Ramalingam, G. (ed.) APLAS 2008. LNCS, vol. 5356, pp. 139–154. Springer, Heidelberg (2008)
14. Ducasse, S., Nierstrasz, O., Schärli, N., Wuyts, R., Black, A.P.: Traits: a mechanism for fine-grained reuse. ACM Trans. Program. Lang. Syst. **28**(2), 331–388 (2006)
15. Felleisen, M., Hieb, R.: The revised report on the syntactic theories of sequential control and state. Theoret. Comput. Sci. **103**(2), 235–271 (1992)
16. Gosling, J., Joy, B., Steele, G., Bracha, G.: The Java(TM) Language Specification, 3rd edn. Addison-Wesley Professional, Reading (2005)
17. Haller, P., Odersky, M.: Scala actors: unifying thread-based and event-based programming. Theoret. Comput. Sci. **410**(2–3), 202–220 (2009)
18. Halstead Jr, R.H.: Multilisp: a language for concurrent symbolic computation. ACM Trans. Program. Lang. Syst. **7**(4), 501–538 (1985)
19. Henrio, L., Huet, F., István, Z.: Multi-threaded active objects. In: De Nicola, R., Julien, C. (eds.) COORDINATION 2013. LNCS, vol. 7890, pp. 90–104. Springer, Heidelberg (2013)
20. Johnsen, E.B., Hähnle, R., Schäfer, J., Schlatte, R., Steffen, M.: ABS: a core language for abstract behavioral specification. In: Aichernig, B.K., de Boer, F.S., Bonsangue, M.M. (eds.) Formal Methods for Components and Objects. LNCS, vol. 6957, pp. 142–164. Springer, Heidelberg (2011)
21. Johnsen, E.B., Owe, O.: An asynchronous communication model for distributed concurrent objects. Softw. Syst. Model. **6**(1), 35–58 (2007)
22. Kitchin, D., Quark, A., Cook, W., Misra, J.: The Orc programming language. In: Lee, D., Lopes, A., Poetzsch-Heffter, A. (eds.) FMOODS 2009. LNCS, vol. 5522, pp. 1–25. Springer, Heidelberg (2009)
23. Lavender, R.G., Schmidt, D.C.: Pattern Languages of Program Design 2. Chapter Active Object: An Object Behavioral Pattern for Concurrent Programming. Addison-Wesley Longman Publishing Co., Inc, Boston (1996)
24. Liskov, B.H., Shrira, L.: Promises: Linguistic support for efficient asynchronous procedure calls in distributed systems. In: Wise, D.S. (ed.) Proceedings of the SIGPLAN Conference on Programming Lanugage Design and Implementation (PLDI 1988), pp. 260–267. ACM, Atlanta, GE, USA (1988)
25. Meseguer, J.: Conditional rewriting logic as a unified model of concurrency. Theoret. Comput. Sci. **96**, 73–155 (1992)
26. Miller, H., Haller, P., Odersky, M.: Spores: a type-based foundation for closures in the age of concurrency and distribution. In: Jones, R. (ed.) ECOOP 2014. LNCS, vol. 8586, pp. 308–333. Springer, Heidelberg (2014)
27. Miller, M.S.: Robust composition: towards a unified approach to access control and concurrency control. Ph.D. thesis, Johns Hopkins University, Baltimore, Maryland, USA, May 2006
28. Nielson, F., Nielson, H.R., Hankin, C.: Principles of Program Analysis. Springer, Heidelberg (1999)
29. Noble, J., Clarke, D.G., Potter, J.: Object ownership for dynamic alias protection. In: TOOLS Pacific 1999: 32nd International Conference on Technology of Object-Oriented Languages and Systems, pp. 176–187. IEEE Computer Society, Melbourne, Australia, 22–25 November 1999

30. Peyton Jones, S., et al.: The Haskell 98 language and libraries: the revised report. J. Funct. Program. **13**(1), 0–255 (2003)

31. Prokopec, A., Bagwell, P., Rompf, T., Odersky, M.: A Generic parallel collection framework. In: Jeannot, E., Namyst, R., Roman, J. (eds.) Euro-Par 2011, Part II. LNCS, vol. 6853, pp. 136–147. Springer, Heidelberg (2011)

32. Reppy, J.H.: Concurrent Programming in ML. Cambridge University Press, Cambridge (1999)

33. Rossberg, A., Botlan, D.L., Tack, G., Brunklaus, T., Smolka, G.: Alice Through the Looking Glass, Munich, Germany. Trends in Functional Programming, vol. 5, pp. 79–96. Intellect Books, Bristol (2006). ISBN 1-84150144-1

34. Srinivasan, S., Mycroft, A.: Kilim: Isolation-typed actors for Java. In: Vitek, J. (ed.) ECOOP 2008. LNCS, vol. 5142, pp. 104–128. Springer, Heidelberg (2008)

35. The Akka Project. Akka (2015). http://akka.io/

36. Van Roy, P., Haridi, S.: Concepts, Techniques, and Models of Computer Programming. MIT Press, Cambridge (2004)

37. Welc, A., Jagannathan, S., Hosking, A.: Safe futures for Java. In: Proceedings of the Object Oriented Programming, Systems, Languages, and Applications (OOPSLA 2005), pp. 439–453. ACM Press, New York, NY, USA (2005)

38. Wrigstad, T., Pizlo, F., Meawad, F., Zhao, L., Vitek, J.: Loci: simple thread-locality for Java. In: Drossopoulou, S. (ed.) ECOOP 2009. LNCS, vol. 5653, pp. 445–469. Springer, Heidelberg (2009)

39. Yokote, Y., Tokoro, M.: Concurrent programming in ConcurrentSmalltalk. In: Yonezawa, A., Tokoro, M. (eds.) Object-Oriented Concurrent Programming, pp. 129–158. The MIT Press, Cambridge, Mass. (1987)

40. Yonezawa, A.: ABCL: An Object-Oriented Concurrent System. Series in Computer Systems. The MIT Press, Cambridge (1990)

41. Yonezawa, A., Briot, J.-P., Shibayama, E:. Object-oriented concurrent programming in ABCL/1. In: Conference on Object-Oriented Programming Systems, Languages and Applications (OOPSLA 1986) (1986). Sigplan Not. 21(11):258–268 (1986)

Coordinating Multicore Computing

Farhad Arbab[1,2]($\boxtimes$) and Sung-Shik T.Q. Jongmans[1]

[1] Formal Methods, CWI, Science Park 123, 1098 XG Amsterdam, The Netherlands
farhad@cwi.nl
[2] Leiden Institute for Advanced Computer Science, Leiden University,
Niels Bohrweg 1, 2333 CA Leiden, The Netherlands

Abstract. Traditional models of concurrency resort to peculiarly indirect means to express interaction and study its properties. Formalisms such as process algebras/calculi, concurrent objects, actors, shared memory, message passing, etc., all are primarily action-based models that provide constructs for the direct specification of *things that interact*, rather than a direct specification of *interaction* (protocols). Consequently, interaction in these formalisms becomes a derived or secondary concept whose properties can be studied only indirectly, as the side-effects of the (intended or coincidental) couplings or clashes of the *actions* whose compositions comprise a model.

Treating interaction as an explicit first-class concept, complete with its own composition operators, allows to specify more complex interaction protocols by combining simpler, and eventually primitive, protocols. Reo [4,7,8,15] serves as a premier example of such an interaction-based model of concurrency. In this paper, we describe Reo and its compiler. We show how exogenous coordination in Reo reflects an interaction-centric model of concurrency where an interaction (protocol) consists of nothing but a relational constraint on communication actions. In this setting, interaction protocols become explicit, concrete, tangible (software) constructs that can be specified, verified, composed, and reused, independently of the actors that they may engage in disparate applications.

This paper complements the first author's lecture at the 15th *International School on Formal Methods for the Design of Computer, Communication and Software Systems* in Bertinoro, Italy, June 2015, and collects previously published material (notably [9]).

1 Introduction

With the availability of today's low-cost multicore commodity hardware that can scale up to offer massively parallel computing platforms, high-speed communication networks that interconnect the globe, plus every indication that both of these phenomena constitute trends that will continue in the future, the need for programming techniques to harness the massive concurrency that they offer has become more vivid than ever. Concurrency is inherently difficult because it involves complex interaction protocols. The inadequacy of traditional models for programming of concurrent systems to serve this purpose stems from the fact that the way in which they express interaction protocols generally does not scale up.

© Springer International Publishing Switzerland 2015
M. Bernardo and E.B. Johnsen (Eds.): SFM 2015, LNCS 9104, pp. 57–96, 2015.
DOI: 10.1007/978-3-319-18941-3_2

Global Objects:

```
1 Semaphore greenSemaphore = new Semaphore(1);
2 Semaphore redSemaphore = new Semaphore(0);
3 Semaphore bufferSemaphore = new Semaphore(1);
4 String buffer = EMPTY;
```

Green Producer:

```
14 while (true) {
15   sleep(5000);
16   greenText = ...;
17   greenSemaphore.acquire();
18   bufferSemaphore.acquire();
19   buffer = greenText;
20   bufferSemaphore.release();
21   redSemaphore.release();
22 }
```

Consumer:

```
5 while (true) {
6   sleep(4000);
7   bufferSemaphore.acquire();
8   if (buffer != EMPTY) {
9     println(buffer);
10    buffer = EMPTY;
11  }
12  bufferSemaphore.release();
13 }
```

Red Producer:

```
23 while (true) {
24   sleep(3000);
25   redText = ...;
26   redSemaphore.acquire();
27   bufferSemaphore.acquire();
28   buffer = redText;
29   bufferSemaphore.release();
30   greenSemaphore.release();
31 }
```

Fig. 1. Alternating producers and consumer

In spite of the fact that *interaction* constitutes the most challenging aspect of concurrency, traditional models of concurrency predominantly treat interaction as a secondary or derived concept. Shared memory, message passing, calculi such as CSP [40], CCS [68], the π-calculus [69,72], process algebras [19,27,36], and the actor model [6] represent popular approaches to tackle the complexities of constructing concurrent systems. Beneath their significant differences, all these models share one common characteristic, inherited from the world of sequential programming: they all constitute *action*-based models of concurrency.

For example, consider developing a simple concurrent application with two producers, which we designate as Green and Red, and one consumer. The consumer must repeatedly obtain and display the contents alternately made available by the Green and the Red producers.

Figure 1 shows the pseudo code for an implementation of this simple application in a Java-like language. Lines 1–4 in this code declare four globally shared entities: three semaphores and a buffer. The semaphores `greenSemaphore` and `redSemaphore` are used by their respective Green and Red producers for their turn keeping. The semaphore `bufferSemaphore` is used as a mutual exclusion lock for the producers and the consumer to access the shared `buffer`, which is initialized to contain the empty string. The rest of the code defines three processes: two producers and a consumer.

The consumer code (lines 5–13) consists of an infinite loop where in each iteration, it performs some computation (which we abstract as the `sleep` on line 6), then it waits to acquire exclusive access to the buffer (line 7). While it has this exclusive access (lines 8–11), it checks to see if the buffer is empty. An empty buffer means there is no (new) content for the consumer process to display, in which case the consumer does nothing and releases the buffer lock (line 12).

If the buffer is non-empty, the consumer prints its content and resets the buffer to empty (lines 9–10).

The Green producer code (lines 14–22) consists of an infinite loop where in each iteration, it performs some computation and assigns the value it wishes to produce to local variable `greenText` (lines 14–15), and waits for its turn by attempting to acquire `greenSsemaphore` (line 17). Next, it waits to gain exclusive access to the shared buffer, and while it has this exclusive access, it assigns `greenText` into `buffer` (lines 18–20). Having completed its turn, the Green producer now releases `redSemaphore` to allow the Red producer to have its turn (line 21).

The Red producer code (lines 23–31) is analogous to that of the Green producer, with "red" and "green" swapped.

This is a simple concurrent application whose code has been made even simpler by abstracting away its computation and declarations. Apart from their trivial outer infinite loops, each process consists of a short piece of sequential code, with a straight-line control flow that involves no inner loops or non-trivial branching. The protocol embodied in this application, as described in our problem statement, above, is also quite simple. One expects it be easy, then, to answer a number of questions about what specific parts of this code manifest the various properties of our application. For instance, consider the following questions:

1. Where is the green text computed?
2. Where is the red text computed?
3. Where is the text printed?

The answers to these questions are indeed simple and concrete: lines 16, 25, and 9, respectively. Indeed, the "computation" aspect of an application typically correspond to coherently identifiable passages of code. However, the perfectly legitimate question "Where is the protocol of this application?" does not have such an easy answer: the protocol of this application is intertwined with its computation code. More refined questions about specific aspects of the protocol have more concrete answers:

1. What determines which producer goes first?
2. What ensures that the producers alternate?
3. What provides protection for the global shared buffer?

The answer to the first question, above, is the collective semantics behind lines 1, 2, 17, and 26. The answer to the second question is the collective semantics behind lines 1, 2, 17, 26, 21, and 30. The answer to the third question is the collective semantics of lines 3, 18, 20, 27, and 29. These questions can be answered by pointing to fragments of code scattered among and intertwined with the computation of several processes in the application. It is far more difficult to identify other aspects of the protocol, such as possibilities for deadlock or livelock, with concrete code fragments. While both concurrency-coordinating actions and computation actions are concrete and explicit in this code, the interaction protocol that they induce is implicit, nebulous, and intangible. In applications involving processes with even slightly less trivial control flow, the entanglement

```
Green Producer:                          Red Producer:
14 while (true) {                        28 while (true) {
15   sleep(5000);                        29   sleep(3000);
16   greenText = ...;                    30   redText = ...;
17   greenSemaphore.acquire();           31   redSemaphore.acquire();
18   while (greenText !=EMPTY) {          32   while (redText !=EMPTY) {
19     bufferSemaphore.acquire();         33     bufferSemaphore.acquire();
20     if (buffer == EMPTY) {             34     if (buffer == EMPTY) {
21       buffer = greenText;              35       buffer = redText;
22       greenText = EMPTY;               36       redText = EMPTY;
23     }                                  37     }
24     bufferSemaphore.release();         38     bufferSemaphore.release();
25   }                                    39   }
26   redSemaphore.release();             40   greenSemaphore.release();
27 }                                      41 }
```

Fig. 2. Busy waiting consumer

of data and control flow with concurrency-coordination actions makes it difficult to determine which parts of the code give rise to even the simplest aspects of their interaction protocol.

When the protocol in a typical concurrent application consists of 623 send and receive (or lock/unlock, etc.) primitives, sprinkled over 783,961 lines of C code, chopped up into 387 different source files, how simple is it to understand this protocol, reason about its properties, debug it, adapt it, or imagine reusing it in another application? How can a hapless programmer (who may very well be the original author of the code, six months down the road) even *see* what this protocol actually does before he can contemplate to do anything with it? Even in the case of our simple program in Fig. 1, which we just examined, do we see all of its properties? We asked about and identified the buffer protection mechanism in this application. But does this mechanism provide adequate protection that we expect?

It is only tactful to say that we are sure all our readers have already spotted what may be considered a bug in this code that may in fact remain undetected in practice for a very long time, depending on the circumstances that determine the relative speeds of the producer and consumer threads in this application. There is no protection in this code preventing the producers from over-writing each other in the buffer, regardless of whether or not their output has actually been consumed by the consumer. Strictly speaking, the original statement of our requirements does not forbid this behavior, so whether this is a bug (in the specification or implementation) is unclear. Suppose the intention in fact was for the consumer to alternately consume what the two producers produce, which means the implementation in Fig. 1 is incorrect and we need to alter it.

One solution is to make the producers sensitive to the emptiness of the buffer. The code for the new producers appears in Fig. 2. A disadvantage of this code is that it more heavily uses the busy-waiting mechanism that already existed in the consumer code in Fig. 1. A better alternative is to use a different protocol that explicitly respects the turn taking, as described below.

In the program shown in Fig. 3, the consumer too has its own turn-taking semaphore, the new **blueSemaphore** (line 3), which is initialized to be locked,

Global Objects:

```
1 Semaphore greenSemaphore = new Semaphore(1);
2 Semaphore redSemaphore = new Semaphore(0);
3 Semaphore blueSemaphore = new Semaphore(0);
4 Semaphore bufferSemaphore = new Semaphore(0);
5 String buffer = EMPTY;
```

Green Producer:

```
12 while (true) {
13   sleep(5000);
14   greenText = ...;
15   greenSemaphore.acquire();
16   buffer = greenText;
17   blueSemaphore.release();
18   bufferSemaphore.acquire();
19   redSemaphore.release();
20 }
```

Consumer:

```
 6 while (true) {
 7   sleep(4000);
 8   blueSemaphore.acquire();
 9   println(buffer);
10   bufferSemaphore.release();
11 }
```

Red Producer:

```
21 while (true) {
22   sleep(3000);
23   redText = ...;
24   redSemaphore.acquire();
25   buffer = redText;
26   blueSemaphore.release();
27   bufferSemaphore.acquire();
28   greenSemaphore.release();
29 }
```

Fig. 3. Revised alternating producers and consumer

just as the redSemaphore, because initially, there is nothing for the consumer to do before any of the producers produces something. The initialization of the bufferSemaphore is also changed (line 4), making the buffer initially locked on behalf of the first producer. The consumer and the two producers all can proceed until each reaches its own turn-taking lock on lines 8, 15, and 24, respectively. The consumer and the Red producer suspend themselves on their turn-taking locks, but the Green producer can proceed beyond its turn-taking lock (line 15), where it fills the buffer (line 16), releases the turn-taking lock of the consumer (line 17), and suspends itself on the buffer lock (line 18). Only the consumer can now proceed, printing the content of the buffer (line 9), and releasing the buffer lock (line 10), after which it proceeds with its next iteration in which it suspends itself on its turn-taking lock (line 8). Only the Green producer can now proceed, having obtained the buffer lock. It now completes its iteration by releasing the turn-taking lock of the Red producer (line 19), and starts its next iteration in which it suspends itself on its own turn-taking lock (line 15). Now, only the Red producer can proceed to fill the buffer (line 25), release the turn- taking lock of the consumer (line 26), and suspend itself on the buffer lock (line 27). The consumer now goes through another iteration, at the end of which it releases the buffer lock, allowing only the Red producer to proceed. The Red producer now releases the turn-taking lock of the Green producer (line 29), and starts its next iteration in which it suspends itself on its own turn-taking lock (line 24) again.

Now that we have a correct protocol (if we indeed do) that does what we expect it to do (if it indeed does), what can we do with this protocol? How easy is it, for instance to reuse this same protocol in a more elaborate application where the control flow of the processes is more complex than the essentially linear, sequential flow of these simple processes? Is it possible to bundle up this protocol and

```
Global Names:                           Green Producer:
  synchronization-points g, r, b, d       G := genG(t).?g(k).!b(t).?d(j).!r(k).G

Consumer:                               Red Producer:
  B := ?b(t).print(t).!d("done").B       R := genR(t).?r(k).!b(t).?d(j).!g(k).R

Application:
  G | R | B | !g("token")
```

Fig. 4. Alternating producers and consumer in a process algebra

parameterize it such that we can instantiate the protocol with arbitrary numbers of processes containing arbitrary computation code, the same way that we can package a piece of code into a parameterized function to compute the inverse of a matrix of any size, or find the minimum element in a list of any size? It would certainly help in software development for multicore platforms, for instance, if we could simply specify the desired numbers of participants and the specific computation code for each, to instantiate an abstract parameterized protocol, as easily as passing arguments in a function call, to tailor the desired concurrency on the available cores. How easy is it to alter this protocol to change the imposed ordering or to allow a pair of considerably fast producers go as fast as they wish, while the slower consumer merely *samples* their output? Such manipulations are difficult with this and similar incarnations of a protocol because they require *seeing* and *touching* the protocol as a tangible concrete entity.

Process algebraic models of concurrency fare only slightly better in this regard than, e.g., programming with threads: they too embody an action-based model of concurrency. Figure 4 shows a process algebraic model of our alternating producers and consumer application. This model consists of a number of globally shared names, i.e., g, r, b, and d. Generally, these shared names are considered as abstractions of channels and thus are called "channels" in the process algebra/calculi community. However, since these names in fact serve no purpose other than synchronizing the I/O operations performed on them, and because we will later use the term "channel" to refer to entities with more elaborate behavior, we use the term "synchronization points" here to refer to "process algebra channels" to avoid confusion.

A process algebra consists of a set of atomic actions, and a set of composition operators on these actions. In our case, the atomic actions include the primitive actions read ?_(_) and write !_(_) defined by the algebra, plus the user-defined actions genG(_), genR(_), and print(_), which abstract away computation. Typical composition operators include sequential composition _ . _, parallel composition _ | _, nondeterministic choice _ + _, definition _ := _, and implicit recursion.

In our model, the consumer B waits to read a data item into t by synchronizing on the global name b, and then proceeds to print t (to display it). It then writes a token "done" on the synchronization point d, and recurses. The Green producer G first generates a new value in t, then waits for its turn by reading a token value into k from g. It then writes t to b, and waits to obtain an acknowledgment j through d, after which it writes the token k to r, and recurses. The Red producer R behaves similarly, with the roles of r and g swapped. The application consists of a parallel composition of the two producers and the consumer, plus a trivial process that simply writes a "token" on g to kick off process G to go first.

Observe that a model is constructed by composing (atomic) actions into (more complex) actions, called processes. True to their moniker, such formalisms are indeed *algebras of processes* or actions. Just as in the version in Fig. 3, while communication actions are concrete and explicit in the incarnation of our application in Fig. 4, *interaction* is a manifestation of the model with no direct explicit structural correspondence. Process algebraic incarnations of concurrency protocols are obviously simpler and more concise than their incarnations in typical programming languages, primarily because they abstract away the clutter of computation. Nevertheless, process algebras and calculi also constitute action-based models of concurrency.

In all action-based models of concurrency, interaction becomes a by-product of processes executing their respective actions: when a process A happens to execute its ith communication action a_i on a synchronization point, at the same time that another process B happens to execute its jth communication action b_j on that same synchronization point, the actions a_i and b_j "collide" with one another and their collision yields an interaction. Manifested this way, an interaction protocol consists of a desired temporal sequence of such (coincidental or planned) collisions. It is non-trivial to distinguish between the essential and the coincidental collision sequences, when the protocol itself is only such an ephemeral manifestation.

Generally, the reason behind the specific collision of a_i and b_j remains debatable. Perhaps it was just dumb luck. Perhaps it was divine intervention. Some may prefer to attribute it to intelligent design! What is not debatable is the fact that, a split second earlier or later, perhaps in another run of the same application, completely random cosmic rays may zap a memory bit and trigger the automatic hardware error correction of the affected memory cell, and thus change the relative timing of the running processes, making a_i and b_j collide not with each other, but with two other actions (of perhaps other processes) yielding completely different interactions. Action based models of concurrency make protocols more difficult than necessary to specify, manipulate, verify, debug, and next to impossible to reuse.

Instead of explicitly composing (communication) actions to indirectly specify and manipulate implicit interactions, is it possible to devise a model of concurrency where interaction (not action) is an explicit, first-class construct? We tend to this question in the next section and in the remainder of this paper describe a specific language based on an interaction-centric model of concurrency. We show that making interaction explicit leads to a clean separation of computation and communication, and produces reusable, tangible protocols that can be constructed and verified independently of the processes that they engage.

2   Interaction-Centric Concurrency

The most salient characteristic of *interaction* is that it transpires among two or more actors. This is in contrast to *action*, which is what a single actor manifests.

In other words, interaction is not about the specific actions of individual actors, but about the relations that (must) hold among those actions. A model of inter-action, thus, must allow us to directly specify, represent, construct, compose, decompose, analyze, and reason about those relations that define what tran-spires among two or more engaged actors, without the necessity to be specific about their individual actions. Making interaction a first-class concept means that a model must offer (1) an explicit, direct, concrete representation of the interaction among actors, independent of their (communication) actions; (2) a set of primitive interactions; and (3) composition operators to combine (primi-tive) interactions into more complex interactions.

Wegner has proposed to consider coordination as constrained interaction [74]. We propose to go a step further and consider interaction itself as a constraint on (communication) actions. Features of a system that involve several entities, for instance the clearance between two physical objects, cannot conveniently be associated with any one of those entities. It is quite natural to specify and rep-resent such features as *constraints*. The interaction among several active entities has a similar essence: although it involves them, it does not *belong* to any one of those active entities. Constraints have a natural formal model as mathematical relations, which are non-directional. In contrast, actions correspond to functions or mappings which are directional, i.e., transformational.

A constraint declaratively specifies *what* must hold in terms of a relation. Typically, there are many ways in which a constraint can be enforced or violated, leading to many different sequences of actions that describe precisely *how* to enforce or maintain a constraint. Action-based models of concurrency lead to the precise specification of *how* in terms of sequences of actions interspersed among the active entities involved in a protocol. In an interaction-based model of concurrency, only *what* a protocol represents is specified as a constraint over the (communication) actions of some active entities; as in constraint programming, the responsibility of how the protocol constraints are enforced or maintained is relegated to an entity other than those active entities.

Generally, composing the sequences of actions that manifest two different protocols does not yield a sequence of actions that manifests a composition of those protocols. Thus, in action-based models of concurrency, protocols are not compositional. Represented as constraints, in an interaction-based model of concurrency, protocols can be composed as mathematical relations.

Banishing the actions that comprise protocol fragments out of the bodies of processes produces simpler, cleaner, and more reusable processes. Expressed as constraints, pure protocols become first-class, tangible, reusable constructs in their own right. As concrete software constructs, such protocols can be embodied into architecturally meaningful *connectors*.

In this setting, a process (or thread, component, service, actor, agent, etc.) offers no methods, functions, or procedures for other entities to call, and it makes no such calls itself. Moreover, processes cannot exchange messages through targeted send and receive actions. In fact, a process cannot refer to any foreign entity, such as another process, the mailbox or message queue of another process, shared variables, semaphores, locks, etc. The only means of communication of

Fig. 5. Protocol in a connector

a process with its outside world is through *blocking I/O operations* that it may perform exclusively on its own *ports*, producing and consuming passive data. A port is a construct analogous to a file descriptor in a Unix process, except that a port is unidirectional, has no buffer, and supports blocking I/O exclusively.

If i is an input port of a process, there are only two operations that the process can perform on i: (1) blocking input get(i, v) waits indefinitely or until it succeeds to obtain a value through i and assigns it to variable v; and (2) input with time-out get(i, v, t) behaves similarly, except that it unblocks and returns false if the specified time-out t expires before it obtains a value to assign to v. Analogously, if o is an output port of a process, there are only two operations that the process can perform on o: (1) blocking output put(o, v) waits indefinitely or until it succeeds to dispense the value in variable v through o; and (2) output with time-out put(o, v, t) behaves similarly, except that it unblocks and returns false if the specified time-out t expires before it dispenses the value in v.

Inter-process communication is possible only by mediation of connectors. For instance, Fig. 5 shows a producer, P and a consumer C whose communication is coordinated by a simple connector. The producer P consists of an infinite loop in each iteration of which it computes a new value and writes it to its local output port (shown as a small circle on the boundary of its box in the figure) by performing a blocking put operation. Analogously, the consumer C consists of an infinite loop in each iteration of which it performs a blocking get operation on its own local input port, and then uses the obtained value. Observe that, written in an imperative programming language, the code for P and C is substantially simpler than the code for the Green/Red producers and the consumer in Figs. 1, 2, and 3: it contains no semaphore operations or any other inter-process communication primitives.

The direction of the connector arrow in Fig. 5 suggests the direction of the dataflow from P to C. However, even in the case of this very simple example, the precise behavior of the system crucially depends on the specific protocol that this simple connector implements. For instance, if the connector implements a synchronous protocol, then it forces P and C to iterate in lock-step, by synchronizing their respective put and get operations in each iteration. On the other hand the connector may have a bounded or an unbounded buffer and implement an asynchronous protocol, allowing P to produce faster than C can consume. The protocol of the connector may, for instance enable it to replicate data items, e.g., the last value that it contained, if C consumes faster and drains the buffer. The protocol may mandate an ordering other than FIFO on the contents of the connector buffer, perhaps depending on the contents of the exchanged data. It may retain only some of the contents of the buffer (e.g., only the first or the last item) if P produces data faster than C can consume. It may be unreliable and lose data nondeterministically or according to some probability distribution. It may retain data in its buffer only for a specified length of time, losing all data items that are not consumed before their expiration dates. The alternatives for the connector protocol are endless, and composed with the very same P and C, each yields a totally different system.

A number of key observation about this simple example are worth noting. First, Fig. 5 is an architecturally informative representation of this system. Second, banishing all inter-process communication out of the communicating parties, into the connector, yields a "good" system design with the beneficial consequences that:

- changing P, C, or the connector does not affect the other parts of the system;
- although they are engaged in a communication with each other, P and C are oblivious to each other, as well as to the actual protocol that enables their communication;
- the protocol embodied in the connector is oblivious to P and C.

In this architecture, the composition of the components and the coordination of their interactions are accomplished *exogenously*, i.e., from outside of the components (or processes) themselves, and without their "knowledge"[1]. In contrast, the interaction protocol and coordination in the examples in Figs. 1, 2, 3, and 4 are *endogenous*, i.e., accomplished through (inter-process communication) primitives from inside the parties engaged in the protocol. It is clear that exogenous composition and coordination lead to simpler, cleaner, and more reusable component code, simply because all composition and coordination concerns are left out. What is perhaps less obvious is that exogenous coordination also leads to reusable, pure coordination code: there is nothing in any incarnation of the connector in Fig. 5 that is specific to P or C; it can just as readily engage any producer and consumer processes in any other application.

Obviously, we are not interested in only this example, nor exclusively in connectors that implement exogenous coordination between only two communicating parties. Moreover, the code for any version of the connector in Fig. 5, or any other connector, can be written in any programming language: the concepts of exogenous composition, exogenous coordination, and the system design and architecture that they induce constitute what matters, not the implementation language.

Nevertheless, focusing on multi-party interaction/coordination protocols reveals that they are composed out of a small set of common recurring concepts. They include synchrony, atomicity, asynchrony, ordering, exclusion, grouping, selection, etc. Encoding every instance of these recurring concepts in terms of assignment statements, if-then-else, for-loops, and communication actions in every application is tedious, error prone, and obscures the concepts beyond recognition when they are interspersed with the computation code of an application. Compliant with the constraint view of interaction advocated above, these concepts can be expressed more succinctly and elegantly as constraints. This observation behooves us to consider the interaction-as-constraint view of concurrency as a foundation for a special language to specify multi-party exogenous interaction/coordination

---

[1] By this anthropomorphic expression we simply mean that a component does not contain any piece of code that directly contributes to determine the entities that it composes with, or the specific protocol that coordinates its own interactions with them.

protocols and the connectors that embody them, of which the connector in Fig. 5 is but a trivial example. Reo, described in the next section, is a premier example of such a language.

3 Overview of Reo

Reo [4,7,8,15] is a channel-based exogenous coordination language wherein complex coordinators, called connectors, or *circuits*, are compositionally built out of simpler ones. Exogenous coordination imposes a purely local interpretation on each inter-components communication, engaged in as a pure I/O operation on each side, that allows components to communicate anonymously, through the exchange of untargeted passive data. We summarize only the main concepts in Reo here. Further details about Reo and its semantics can be found in the cited references.

Complex connectors in Reo are constructed as a network of primitive binary connectors, called *channels*. Connectors serve to provide the protocol that controls and organizes the communication, synchronization and cooperation among the components/services that they interconnect. Formally, the protocol embodied in a connector is a *relation*, which the connector imposes as a *constraint* on the actions of the communicating parties that it inter-connects.

A channel is a medium of communication that consists of two ends and a constraint on the dataflows observed at those ends. There are two types of channel ends: *source* and *sink*. A source channel end accepts data into its channel, and a sink channel end dispenses data out of its channel. Every channel (type) specifies its own particular behavior as constraints on the flow of data through its ends. These constraints relate, for example, the content, the conditions for loss, and/or creation of data that pass through the ends of a channel, as well as the atomicity, exclusion, order, and/or timing of their passage. Reo places no restriction on the behavior of a channel and thus allows an open-ended set of different channel types to be used simultaneously together.

Although all channels used in Reo are user-defined and users can indeed define channels with any complex behavior (expressible in a semantic model of Reo) that they wish, a very small set of channels, each with very simple behavior, suffices to construct useful Reo connectors with significantly complex behavior. Figure 6 shows a common set of primitive channels often used to build Reo connectors.

| Sync | LossySync | FIFO1 | SyncDrain | AsyncDrain | Filter(P) |

Fig. 6. A typical set of Reo channels

A Sync channel has a source and a sink end and no buffer. It accepts a data item through its source end iff it can simultaneously (i.e., atomically) dispense it through its sink.

A `LossySync` channel is similar to a synchronous channel except that it always accepts all data items through its source end. This channel transfers a data item if it is possible for the channel to dispense the data item through its sink end; otherwise the channel loses the data item. Observe that the behavior of this channel if fully deterministic; the channel is never free to choose between passing or losing a data item: if it is possible for a data item to be consumed through its sink end, the channel *must* pass the data item exactly as a `Sync`. Thus, the context of (un)availability of a ready consumer at its sink end determines the (context-sensitive) behavior a `LossySync` channel.

A `FIFO1` channel represents an asynchronous channel with a buffer of capacity 1: it can contain at most one data item. In the graphical representation of an empty `FIFO1` channel, no data item is shown in the box (this is the case in Fig. 1). If the buffer of a `FIFO1` channel contains a data element d, then d appears inside the box in its graphical representation. When its buffer is empty, a `FIFO1` channel blocks I/O operations on its sink, because it has no data to dispense. It dispenses a data item and allows an I/O operation at its sink to succeed, only when its buffer is full, after which its buffer becomes empty. When its buffer is full, a `FIFO1` channel blocks I/O operations on its source, because it has no more capacity to accept the incoming data. It accepts a data item and allows an I/O operation at its source to succeed, only when its buffer is empty, after which its buffer becomes full.

More exotic channels are also permitted in Reo, for instance, synchronous and asynchronous *drains*. Each of these channels has two source ends and no sink end. No data value can be obtained from a drain channel because it has no sink end. Consequently, all data accepted by a drain channel are lost. `SyncDrain` is a synchronous drain that can accept a data item through one of its ends iff a data item is also available for it to simultaneously accept through its other end as well. `AsyncDrain` is an asynchronous drain that accepts data items through its source ends and loses them exclusively one at a time, but never simultaneously.

For a *filter channel*, or `Filter(P)`, its pattern $P \subseteq Data$ specifies the type of data items that can be transmitted through the channel. This channel accepts a value $d \in P$ through its source end iff it can simultaneously dispense d through its sink end, exactly as if it were a `Sync` channel; it always accepts all data items $d \notin P$ through its source end and loses them immediately.

Synchronous and asynchronous *Spouts* are the duals of their respective drain channels, as each has two sink ends through which it produces nondeterministic data items. Further discussion of these and other primitive channels is beyond the scope of this paper.

Complex connectors are constructed by composing simpler ones via the *join* and *hide* operations. Channels are joined together in *nodes*, each of which consists of a set of channel ends. A Reo node is a logical place where channel ends coincide

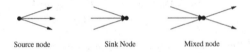

Source node   Sink Node   Mixed node

Fig. 7. Reo nodes

and coordinate their dataflows as prescribed by its *node type*. Figure 7 shows the three possible node types in Reo. A node is either *source*, *sink*, or *mixed*, depending on whether all channel ends that coincide on that node are source ends, sink ends, or a combination of the two. Reo fixes the semantics of (i.e., the constraints on the dataflow through) Reo nodes, as described below. The *hide* operation is used to hide the internal topology of a Reo connector. A hidden node can no longer be accessed or observed from outside.

The source and sink nodes of a connector are collectively called its *boundary nodes*. Boundary nodes define the interface of a connector. Processes (or components, actors, agents, etc.) connect to the boundary nodes of a connector and interact anonymously with each other through this interface. Connecting a process to a (source or sink) node of a connector consists of the identification of one of the (respectively, output or input) ports of the process with that node. At most one process can be connected to a (source or sink) node at a time. Processes interact by performing their blocking I/O operations on their own local ports, which trigger dataflow through their respectively identified nodes of the connector(s): the get and put operations mentioned in the description of the processes in Fig. 5 trigger *write* and *take* operations of Reo on the channel ends of their respective nodes.

A component (or process) can write data items to a source node that it is connected to. The write operation succeeds only if all (source) channel ends coincident on the node accept the data item, in which case the data item is transparently written to every source end coincident on the node. A source node, thus, acts as a synchronous replicator.

A component (or process) can obtain data items, by an input operation, from a sink node that it is connected to. A take operation succeeds only if at least one of the (sink) channel ends coincident on the node offers a suitable data item; if more than one coincident channel end offers suitable data items, one is selected nondeterministically. A sink node, thus, acts as a nondeterministic merger.

A mixed node nondeterministically selects and takes a suitable data item offered by one of its coincident sink channel ends and replicates it into all of its coincident source channel ends. Note that a component cannot connect to, take from, or write to mixed nodes.

Because a node has no buffer, data cannot be stored in a node. Specifically, a mixed node cannot take a data item out of one of its coincident sink channel ends, unless it can atomically replicate and write it into all of its coincident source channel ends. Hence, nodes instigate the propagation of synchrony and exclusion constraints on dataflow throughout a connector. Deriving the semantics of a Reo connector amounts to resolving the composition of the constraints of its constituent channels and nodes [33]. This is not a trivial task. In the sequel, we present examples of Reo connectors that illustrate how non-trivial dataflow behavior emerges from composing simple channels using Reo nodes. The local constraints of individual channels propagate through (the synchronous regions of) a connector to its boundary nodes. This propagation also induces a certain context-awareness in connectors. See [32] for a detailed discussion of this.

Reo has been used for composition of Web services [16,57,65], modeling and analysis of long-running transactions in service-oriented systems [60], coordination of multi-agent systems [10], performance analysis of coordinated compositions [12,13,17,70,71], modeling of business processes and verification of their compliance [14,59,73], and modeling of coordination in biological systems [31].

Reo offers a number of operations to reconfigure and change the topology of a connector at run-time: operations that enable the dynamic creation of channels, splitting and joining of nodes, hiding internal nodes. The hiding of internal nodes allows to permanently fix the topology of a connector, such that only its boundary nodes are visible and available. The resulting connector can then be viewed as a new primitive connector, or primitive for short, since its internal structure is hidden and its behavior is fixed.

Tool support for Reo consists of a set of Eclipse plug-ins that together comprise the Extensible Coordination Tools (ECT) visual programming environment [2]. The Reo graphical editor supports drag-and-drop graphical composition and editing of Reo connectors. This editor also serves as a bridge to other tools, including animation and code generation plug-ins. The animation plug-in automatically generates a graphical animation of the flow of data in a Reo connector, which provides an intuitive insight into their behavior through visualization of how they work. Several model checking tools are available for analyzing Reo. The Vereofy model checker, integrated in ECT, is based on constraint automata [5,21–24,28,37,55,56]. Properties of Reo connectors can be specified for verification by Vereofy in a language based on Linear Temporal Logic (LTL), or on a variant of Computation Tree Logic (CTL), called Alternating-time Stream Logic (ASL). Another means for verification of Reo is made possible by a transformation bridge into the mCRL2 toolset [3,38]. The mCRL2 verifier relies on the parameterized boolean equation system (PBES) solver to encode model checking problems, such as verifying first-order modal-calculus formulas on linear process specifications. An automated tool integrated in ECT translates Reo models into mCRL2 and provides a bridge to its tool set. This translation and its application for the analysis of workflows modeled in Reo are discussed in [58,62,63]. Through mCRL2, it is possible to verify the behavior of timed Reo connectors, or Reo connectors with more elaborate data-dependent behavior than Vereofy supports. The resulting labeled transformation systems can also be used for analysis by a number of tools in the CADP tool set [1]. Another tool is a Reo compiler that generates executable code for Reo connectors; we discuss compilation in more detail shortly. Even more tools are discussed elsewhere [9].

4   Examples

Recall our alternating producers and consumer example of Sect. 1. We revise the code for the Green and Red producers to make them suitable for exogenous coordination (which, in fact, makes them simpler). Similar to the producer P in Fig. 5, this code now consists of an infinite loop, in each iteration of which the producer computes a new value and writes it to its output port. Analogously, we

```
Consumer:                Green Producer:              Red Producer:
 1 while (true) {         6 while (true) {             11 while (true) {
 2   sleep(4000);         7   sleep(5000);             12   sleep(3000);
 3   get(input, text);    8   greenText = ...;         13   redText = ...;
 4   print(text);         9   put(output, greenText);  14   put(output, redText);
 5 }                     10 }                          15 }
```

Fig. 8. Generic reusable producers and consumer

revise the consumer code, fashioning it after the consumer C in Fig. 5. Figure 8 shows this code.

In the remainder of this section, we present a number of protocols to implement different versions of the alternating producers and consumer example of Sect. 1, using the producers and consumer processes in Fig. 8. These examples serve three purposes. First, they show a flavor of programming of pure interaction coordination protocols as Reo connectors. Second, they present a number of generically useful connectors that can serve as connectors in many other applications, or as sub-connectors in the connectors for construction of many other protocols. Third, they illustrate the utility of exogenous coordination by showing how trivial it is to change the protocol of an application, without altering any of the processes involved.

4.1   Alternator

The connector shown in Fig. 9(a) is an *alternator* that imposes an ordering on the flow of the data from its input nodes A and B to its output node C. The SyncDrain enforces that data flow through A and B only synchronously (i.e., atomically). The empty buffer of the FIFO1 channel together with the SyncDrain guarantee that the data item obtained from A is delivered to C while the data item obtained from B is stored in the FIFO1 buffer. After this, the buffer of the FIFO1 is full and data cannot flow in through either A or B, but C can dispense the data stored in the FIFO1 buffer, which makes it empty again. Thus, subsequent take operations at C obtain the data items written to $A, B, A, B, ...,$ etc.

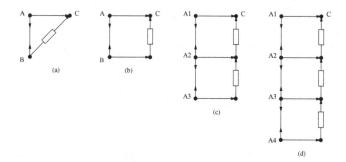

Fig. 9. Reo connectors for alternators

The connector in Fig. 9(b) has an extra Sync channel between node B and the FIFO1 channel, compared to the one in Fig. 9(a). It is trivial to see that these two connectors have the exact same behavior. However, the structure of the connector in Fig. 9(b) allows us to generalize its alternating behavior to any number of producers, simply by replicating it and "juxtaposing" the top and the bottom Sync channels of the resulting copies, as seen in Fig. 9(c) and (d).

The two SyncDrain channels in the connector shown in Fig. 9(c) require data to flow through $A1$, $A2$, and $A3$ only simultaneously (i.e., atomically). The empty buffers of the FIFO1 channels, together with these SyncDrain channels guarantee that the data item obtained from $A1$ is delivered to C while the data items obtained from $A2$ and $A3$ are stored in the buffers of their respective FIFO1 channels. Subsequently, as long as the buffer of at least one of the FIFO1 channels remains full, no data can flow through any of the nodes $A1$, $A2$, and $A3$, but C can dispense the data stored in the buffers of the FIFO1 channels, with their order preserved. Thus, the first 3 take operations on C deliver the data items obtained through $A1$, $A2$, and $A3$, in that order. At this point, all FIFO1 buffers become empty and the next round of input becomes possible.

The connector in Fig. 9(d) is obtained by replicating the one in Fig. 9(b) 3 times. Following the reasoning for the connector in Fig. 9(c), it is easy to see that the connector in Fig. 9(d) delivers the data items obtained from $A1$, $A2$, $A3$,and $A4$ through C, in that order.

A version of our alternating producers and consumer example of Sect. 1 can now be composed by attaching the output port of the revised Green producer in Fig. 8 to node A, the output port of the revised Red producer in Fig. 8 to node B, and the input port of the consumer in Fig. 8 to node C of the Reo connector in Fig. 9(a).

A closer look shows, however, that the behavior of this version of our example is *not* exactly the same as that of the one in Figs. 3 and 4. As explained above, the Reo connector in Fig. 9(a) requires the availability of a pair of values on A (from the Green producer) and B (from the Red producer) before it allows the consumer to obtain them, first from A and then from B. Thus, if the Green producer and the consumer are both ready to communicate, they still have to wait for the Red producer to also attempt to communicate, before they can exchange data. The versions in Figs. 3 and 4 allow the Green producer and the consumer to go ahead, regardless of the state of the Red producer. Our original specification of this example in Sect. 1 was abstract enough to allow both alternatives. A further refinement of this specification may indeed prefer one and disallow the other. If the behavior of the connector in Fig. 9(a) is *not* what we want, we need to construct a different Reo connector to impose the same behavior as in Figs. 3 and 4. This is precisely what we describe below.

4.2 Sequencer

Figure 10(a) shows an implementation of a sequencer by composing five Sync channels and four FIFO1 channels together. The first (leftmost) FIFO1 channel is initialized to have a data item in its buffer, as indicated by the presence of the

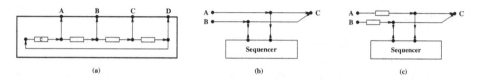

Fig. 10. Sequencer

symbol e in the box representing its buffer cell. The actual value of the data item is irrelevant. The connector provides only the four nodes A, B, C and D for other entities (connectors or component instances) to take from. The take operation on nodes A, B, C and D can succeed only in the strict left-to-right order. This connector implements a generic sequencing protocol: we can parameterize this connector to have as many nodes as we want simply by inserting more (or fewer) Sync and FIFO1 channel pairs, as required.

Figure 10(b) shows a simple example of the utility of the sequencer. The connector in this figure consists of a two-node sequencer, plus a SyncDrain and two Sync channels connecting each of the nodes of the sequencer to the nodes A and C, and B and C, respectively. Similar to the connector in Fig. 9(a), this connector imposes an order on the flow of the data items written to A and B, through C: the sequence of data items obtained by successive take operations on C consists of the first data item written to A, followed by the first data item written to B, followed by the second data item written to A, followed by the second data item written to B, and so on. However, there is a subtle difference between the behavior of the two connectors in Figs. 9(a) and 10(b). The alternator in Fig. 9(a) delays the transfer of a data item from A to C until a data item is also available at B. The connector in Fig. 10(b) transfers from A to C as soon as these nodes can satisfy their respective operations, regardless of the availability of data on B.

We can obtain a new version of our alternating producers and consumer example by attaching the output port of the Green producer in Fig. 8 to node A, the output port of the Red producer in Fig. 8 to node B, and the input port of the consumer in Fig. 8 to node C. The behavior of this version of our application is now the same as the programs in Fig. 4 and in Fig. 1 (after replacing its producers with the ones in Fig. 2). The connector in Fig. 10(b) embodies the same protocol that is implicit in Fig. 4.

A characteristic of this protocol is that it "slows down" each producer, as necessary, by delaying the success of its data production until the consumer is ready to accept its data. Our original problem statement in Sect. 1 does not explicitly specify whether or not this is a required or permissible behavior. While this may be desirable in some applications, slowing down the producers to match the processing speed of the consumer may have serious drawbacks in other applications, e.g., if these processes involve time-sensitive data or operations. Perhaps what we want is to bind our producers and consumer by a protocol that decouples them such as to allow each process to proceed at its own pace. We proceed, below, to present a number of protocols that we then compose to construct a Reo connector for such a protocol.

4.3 Buffered Sequencing

Figure 10(c) shows how easily we can decouple the producers from the consumer by adding two FIFO1 channels to the connector in Fig. 10(b). The protocol implemented by this connector allows each producer to move ahead of its turn by one item. Obviously, one can add more FIFO1 channels, as desired, to allow the producers to move ahead of their turns by any arbitrary k items, before they need to wait for their next output item to be accepted. Because Reo allows users to define arbitrary channels, it is equally possible to define an unbounded FIFO channel, and use two instances of this channel to allow producers to move ahead of the consumer by any arbitrary number of items.

A characteristic of all such buffered protocols is that they make sure every item produced by every producer is eventually consumed by the consumer. In fact, such total retention of data is not always desirable. Sometimes, some sort of *sampling* is required to ensure the consumer is not overwhelmed by much faster producers, or to ensure that the consumer always processes the most up-to-date produced items.

4.4 Sampling

The connector in Fig. 11(a) is a variant of the one in Fig. 10(b) which never delays any of its producers. Producers can produce items as fast as they wish and the protocol never delays them; it simply loses any item that they produce when the consumer is not ready to take it. Whenever the consumer is ready to take an item, it must wait for the producer whose turn it is to produce its next item for it to consume. On the one hand, this ensures that the consumer always obtains the freshest, most up-to-date item produced by each producer. On the other hand, although the producers never wait, the consumer may still have to wait for the right producer to deliver its next fresh item. If this is not desirable, we may wish the protocol to hold at least one produced item at hand to alleviate the need for the consumer to wait.

The connector on the left-hand side of the $\equiv$ sign in Fig. 11(b) shows a useful connector which behaves almost exactly as a FIFO1 channel. The only difference is that, unlike a normal FIFO1 channel, this connector does not suspend its writer if its buffer is full; it allows the write to succeed, but loses the written data. We use the symbol on the right-hand side of the $\equiv$ sign in Fig. 11(b) as a short-hand for this connector, and refer to it as an OverflowLossyFIFO1 channel. This symbol is intentionally similar to that of a regular FIFO1 channel, because the behavior

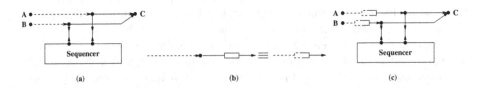

Fig. 11. Synchronized sampling, OverflowLossyFIFO1, and buffered sampling

of this connector closely resembles that of a regular `FIFO1` channel. The dashed source-side half of this channel suggests that when its buffer is full, this channel simply loses its new input items, as if they "overflow" over a full container.

Replacing the `FIFO1` channels in Fig. 10(c) with such `OverflowLossyFIFO1` channels, we obtain the connector in Fig. 11(c). Using this connector in our running example application allows the producers to run as fast as they wish, and allows the consumer to merely sample what each producer delivers. If the consumer ever gets ahead of a producer by more than one cycle, then this protocol makes the consumer wait. Obviously, we can add more `FIFO1` channels to the construct in Fig. 11(b) to obtain an `OverflowLossyFIFO$k$` channel, for any $k > 1$. We can then raise the sampling depth of our protocol to any k by using `OverflowLossyFIFO$k$` channels in connectors similar to the one in Fig. 11(c). Such a protocol with the sampling depth of k allows the consumer move ahead of a producer by k items, while the protocol retains up to k items produced by each producer, before it loses their excess output.

A consequence of using `OverflowLossyFIFO$k$` channels in the above connectors is that the protocol tends to retain the "oldest" k sampled output of each producer.[2] In many situations, it is desirable to bias sampling toward most recent values, discarding older values. To do this, we need a counterpart of the `OverflowLossyFIFO1` channel in Fig. 11(b), that when its buffer is full, discards the old value in the buffer and retains its new input. We present a connector with such behavior in Sect. 4.6.

4.5 Exclusive Router

The connector shown in Fig. 12(a) is a binary *exclusive router*: it routes data from A to either B or C (but not both). This connector can accept data only if there is a write operation at the source node A, and there is at least one taker at the sink node B or C. If both B and C can dispense data, the choice of routing to B or C

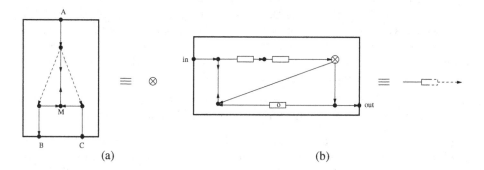

(a)   (b)

Fig. 12. An exclusive router and a `ShiftLossyFIFO1`

---

[2] In fact, this characterization is not very accurate for values of $k > 1$. Work out what happens for $k = 2$, for instance.

follows from the non-deterministic decision by the mixed node M: it can accept
data only from one of its sink ends, excluding the flow of data through the other,
which forces the latter's respective LossySync to lose the data it obtains from A,
while the other LossySync passes its data as if it were a Sync.

By connecting the source node of a binary exclusive router to one of the sink
nodes of another binary exclusive router we obtain a ternary exclusive router.
This is possible in Reo because synchrony and exclusion constraints propagate
through its nodes. More generally, an n-ary exclusive router (with a single source
and n sink ends) can be composed out of $n-1$ binary exclusive routers. Because
the exclusive routers are so commonly useful, we use a graphical short-hand to
represent them in connectors. The crossed circle shown on the right-hand side
of the $\equiv$ symbol in Fig. 12(a) is the symbol that we use to represent a generic
n-ary exclusive router.

4.6   Shift-Lossy FIFO1

Figure 12(b) shows a Reo connector for a connector that behaves as a lossy
FIFO1 channel with a shift loss-policy. This channel is called shift-lossy FIFO1
(ShiftLossyFIFO1). This connector is composed of an exclusive router (shown
in Fig. 12(a)), an initially full FIFO1 channel, two initially empty FIFO1 channels,
and four Sync channels. Intuitively, it behaves as a normal FIFO1 channel, except
that if its buffer is full then the arrival of a new data item deletes the existing
data item in its buffer, making room for the new arrival. As such, this channel
implements a "shift loss-policy" losing the older contents in its buffer in favor of
the newer arrivals. This is in contrast to the behavior of an overflow-lossy FIFO1
channel, whose "overflow loss-policy" loses the new arrivals when its buffer is
full. See [25] for a more formal treatment of the semantics of this connector.

The ShiftLossyFIFO1 connector in Fig. 12(b) is indeed so frequently useful
as a connector in construction of more complex connectors, that it makes sense to
have a special graphical symbol to designate it as a short-hand. The symbol shown
on the right-hand side of the $\equiv$ symbol in Fig. 12(b) is the what we use to represent this connector, and also take the liberty to refer to it as a ShiftLossyFIFO1
channel. This symbol is intentionally similar to that of a regular FIFO1 channel,
because the behavior of this connector closely resembles that of a regular FIFO1
channel. The dashed sink-side half of this channel suggests that it loses the older
values to make room for new arrivals, i.e., it shifts to lose.

4.7   Decoupled Alternating Producers and Consumer

Figure 13(a) shows how the ShiftLossyFIFO1 connector of Fig. 12(b) can be
used to construct a version of the example in Fig. 5, where the producer and the
consumer are partially decoupled from one another. Whenever, as initially is the
case, the ShiftLossyFIFO1 buffer is empty, the consumer has no choice but to
wait for the producer to place a value into this buffer. However, the producer
never has to wait for the consumer: it can work at its own pace and write to the
connector whenever it wishes. Every write by the producer replaces the current

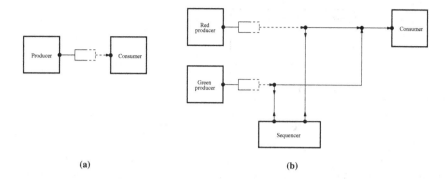

(a)   (b)

Fig. 13. Decoupled producers and consumer

contents of the `ShiftLossyFIFO1` buffer. A subsequent take by the consumer obtains the current value out of `ShiftLossyFIFO1` buffer and makes it empty. The producer never has to wait for the consumer, but if the consumer is faster than the producer, it has to wait for the next data item to arrive. It is instructive to compare the behavior of this system with that of a single `LossySync` channel connecting a producer and a consumer: the two are not exactly the same.

The connector in Fig. 13(b) is a small variation of the Reo connector in Fig. 10(b), with two instances of the `ShiftLossyFIFO1` connector of Fig. 12(b) spliced in. In this version of our alternating producers and consumer, these three processes are partially decoupled: each producer runs at its own pace, never having to wait for any of the other two processes. Every take by the consumer, always obtains and empties the latest value produced by its respective producer. If the consumer runs slower than a producer, the excess data that they produce is lost in the producer's respective `ShiftLossyFIFO1`, which allows the consumer to effectively "sample" the data generated by this producer. If the consumer runs faster than a producer, it will block on its respective empty `ShiftLossyFIFO1` until a new value becomes available.

4.8 Dataflow Variable

The Reo connector in Fig. 14 implements the behavior of a dataflow variable. It uses two instances of the `ShiftLossyFIFO1` connector shown Fig. 12(b), to build a connector with a single input and a single output nodes. Initially, the buffers of its `ShiftLossyFIFO1` channels are empty, so an initial take on its output node suspends for data. Regardless of the status of its buffers, or whether or not data can be dispensed through its output node, every write to its input node always succeeds and resets both of its buffers to contain the new data item. Every time a value is dispensed through its output node, a copy of this value is "cycled back" into its left `ShiftLossyFIFO1` channel. This connector "remembers" the last value it obtains through its input node, and dispenses copies of this value through its output node as frequently as necessary: i.e., it can be used as a dataflow variable.

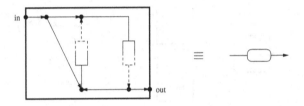

Fig. 14. Dataflow variable

The variable connector in Fig. 14 is also very frequently useful as a connector in construction of more complex connectors. Therefore, it makes sense to have a short-hand graphical symbol to designate it with as well. The symbol shown on the right-hand side of Fig. 14 is the what we use to represent this connector, and also take the liberty to refer to it as a **Variable** channel, or just a "variable" for short. This symbol is intentionally similar to that of a regular **FIFO1** channel, because the behavior of this connector closely resembles that of a regular **FIFO1** channel. We use a rounded box to represent its buffer: the rounded box hints at the recycling behavior of the variable connector, which implements its remembering of the last data item that it obtained or dispensed.

4.9 Fully Decoupled Alternating Producers and Consumer

Figure 15(a) shows how the variable connector of Fig. 14 can be used to construct a version of the example in Fig. 5, where the producer and the consumer are fully decoupled from one another. Initially, the variable contains no value, and therefore, the consumer has no choice but to wait for the producer to place its first value into the variable. After that, neither the producer, nor the consumer ever has to wait for the other one. Each can work at its own pace and write to or take from the connector. Every write by the producer replaces the current contents of the variable, and every take by the consumer obtains a copy of the current value of the variable, which always contains the most recent value produced.

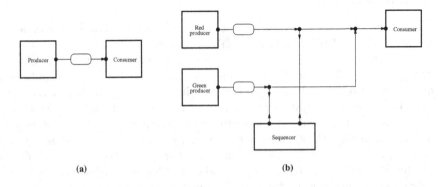

Fig. 15. Fully decoupled producers and consumer

The connector in Fig. 15(b) is a small variation of the Reo connector in Fig. 10(b), with two instances of the variable connector of Fig. 14 spliced in. In this version of our alternating producers and consumer, these three processes are fully decoupled: each can produce and consume at its own pace, never having to wait for any of the other two. Every take by the consumer, always obtains the latest value produced by its respective producer. If the consumer runs slower than a producer, the excess data is lost in the producer's respective variable, and the consumer will effectively "sample" the data generated by this producer. If the consumer runs faster than a producer, it will read (some of) the values of this producer multiple times.

4.10   Flexibility and Scaling

Figures 9(a), 10(b), (c), 11(a), (c), 13(b), and 15(b) show a number of different connectors, each imposing a variant of a protocol for the coordination of two alternating producers and a consumer. The exact same producers and consumer processes can be combined with any of these connectors to yield different applications. It is instructive to compare the ease with which this is accomplished in our interaction-centric world, with the effort involved in modifying the action-centric incarnations of this same example in Figs. 3 and 4, which correspond to the protocol of the connector in Fig. 10(b), in order to achieve the behavior induced by the connector in Figs. 9(a), 10(c), 11(a), (c), 13(b), or 15(b). It is also instructive to compare the ease with which any of these connectors can be parameterized to scale up their number of producers, with the changes necessary to scale up the number of producers in action-centric versions of these protocols.

Moreover, applications with many producers may indeed require somewhat different treatment of the output of some of their producers. For instance, an application may require barrier synchronization of some producers, synchronous sampling of some others, buffered sampling of yet others, etc., etc. It is trivial to mix-and-match the necessary interaction (sub-)protocols that we examined, to tailor make such a protocol, essentially through cut-and-paste of parts of the various Reo connectors presented above. Such cut-and-paste is generally unthinkable when protocols are expressed in terms of action-based constructs of traditional models of concurrency.

As if anyone needed more evidence to appreciate that concurrency is difficult, the many variants of our deceptively trivial running example presented above, plus the multitudes of their possible mix-and-match variants, demonstrate that even seemingly trivial protocols involve intricate details that require careful attention and explicit, concrete, first-class treatment. By the way, none of the variants of the Reo connectors presented above captures the behavior of the Java-like code of our initial attempt. For the sake of completeness, the behavior of the protocol in Fig. 1 corresponds to the behavior of the connector in Fig. 16. Just as in the case of the program in Fig. 1, this connector allows the producers at nodes A and B alternate and over-write each other in the buffer of the ShiftLossyFIFO1. The consumer at C can obtain only the latest value produced by either of the producers.

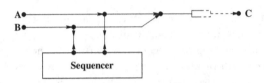

Fig. 16. Alternating and over-writing

The Reo connector binding a number of distributed processes, such as Web services, can even be "hot-swapped" while the application runs, without the knowledge or the involvement of the engaged processes. A prototype platform to demonstrate this capability is available at [2].

5   Semantics

Reo allows arbitrary user-defined channels as primitives; arbitrary mix of synchrony and asynchrony; and relational constraints between input and output. This makes Reo more expressive than, e.g., dataflow models, Kahn networks, synchronous languages, stream processing languages, workflow models, and Petri nets. On the other hand, it makes the semantics of Reo quite non-trivial.

Various models for the formal semantics of Reo have been developed, each to serve some specific purposes. In the rest of this section, we briefly describe constraint automata [25], the main semantics used in verification and code generation; a comprehensive overview of other models appears elsewhere [44].

Constraint automata provide an operational model for the semantics of Reo connectors. The states of an automaton represent the configurations of its corresponding connector (e.g., the contents of the FIFO channels), while the transitions encode its maximally-parallel stepwise behavior. The transitions are labeled with the maximal sets of nodes on which dataflow occurs simultaneously, and a data constraint (i.e., boolean condition for the observed data values). For example, Fig. 17 shows the constraint automata semantics for some of the common Reo primitives.

The constraint automaton for the Sync channel consists of a single state. It has only a single transition, labeled by the pair of *synchronization constraint*, and *data constraint*. The synchronization constraint $\{A, B\}$ states that this transition is possible iff both nodes A and B can *fire* synchronously (i.e., atomically), allowing their respective pending I/O operations to succeed. The data constraint

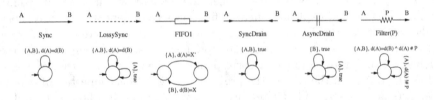

Fig. 17. Constraint automata of some typical Reo channels

$d(A) = d(B)$ states that this transition is possible iff the data observed at node A is identical to the data observed at node B. Because these two nodes are respectively the source and the sink nodes (of the Sync channel), this data constraint requires a transfer of data from A to B.

The constraint automaton for the LossySync channel in fact expresses the semantics of a *nondeterministic* LossySync channel, *not* that of our *context sensitive* LossySync described in Sect. 3. The difference is significant, but it is not important for our purposes in this paper.[3] This automaton has a single state and two transitions. One of these transitions is identical to that of the Sync channel, modeling its identical behavior. The other, labeled by $\{A\}, true$ simply states that the automaton can make this transition iff A can fire by itself and imposes no constraint of the data of A: this data is lost.

The constraint automaton for the FIFO1 channel has two states, representing its empty (initial) and full states. To simplify our presentation, we consider a variant of constraint automata that allow states to have local memory variables. The label $\{A\}, d(A) = X'$ of the transition that takes the automaton from its empty to its full state allows it to make this transition iff node A can fire by itself, and the *new value* of the memory variable X in the target state (identified by X' in the data constraint) is the same as the data value observed on node A: the value obtained from the source node A gets assigned to the X variable of the target state to satisfy this constraint. The label $\{B\}, d(B) = X$ of the transition that takes the automaton from its full to its empty state allows it to make this transition iff node B can fire by itself, and the value of the memory variable X in the source state (identified by X in the data constraint) is the same as the data value observed on node B: the value of the X variable of the source state is dispensed through the sink node B to satisfy this data constraint.

The constraint automaton for the SyncDrain channel has a single state and a single transition, whose constraints require its ends to fire synchronously ($\{A, B\}$), but imposes no constraints ($true$) on their data. Because these are both source ends, their data are simply lost.

The constraint automaton for the AsyncDrain channel has a single state and two transitions, each of which allow it to fire and lose the data obtained through one of its ends (but never both synchronously).

The constraint automaton for the Filter(P) channel has a single state and two transitions. If source node A can fire and its data value does not match the filter pattern P, then the data value of A is simply lost. If the data value available on the source node A matches the filter pattern P, then the only possible transition is one similar to that of the Sync channel, by which the data value of A is transferred to the sink node B.

---

[3] In fact, constraint automata do not have the expressiveness required to directly represent context sensitivity. Other more expressive semantic models, including more sophisticated automata models, have been devised for this purpose [29,34]. A recent work shows that, although constraint automata cannot directly represent context sensitivity, it is possible to *encode* context sensitivity using constraint automata as well [52,61].

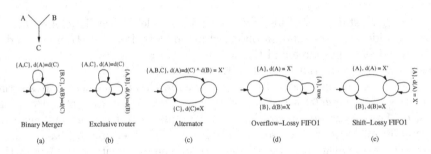

Fig. 18. Constraint automata of a binary merger and some example connectors

The semantics of a Reo connector is derived by composing the constraint automata of its constituents, through a special form of synchronized product of automata, which automatically accommodates the replication semantics of Reo nodes [25]. The nondeterministic n-ary merge semantics inherent in Reo nodes needs to be made explicit as a (product) composition of $n - 1$ nondeterministic binary merge primitives. Figure 18(a) shows the constraint automaton for a nondeterministic binary merge primitive.

Figure 18(b) shows the constraint automaton representing the semantics of the exclusive router Reo connector of Fig. 12(a), which is obtained as the product of the constraint automata of its constituents: 5 Sync channels, 2 LossySync channels, a SyncDrain channel, and a merger.

Figure 18(c) shows the constraint automaton representing the semantics of the alternator connector of Fig. 9(a), obtained as the product of the constraint automata of its constituent Sync channel, SyncDrain channel, FIFO1 channel, and merger.

Figure 18(d) shows the constraint automaton representing the semantics of an *overflow lossy* connector, which can be easily composed by connecting the sink end of a LossySync to the source end of a FIFO1. Although this *is* the semantics that must be obtained, the product of simple constraint automata in Fig. 17 does *not* yield this automaton. This automaton can be obtained using more sophisticated variants of constraint automata [29,34], or an encoding technique [52] which can handle context sensitivity.

Figure 18(e) shows the constraint automaton representing the semantics of the ShiftLossyFIFO1 connector of Fig. 12(b), which is obtained as the product of the constraint automata of its constituents.

Constraint automata have been used for the verification of protocols through model-checking [5,22–24,28,37,55,56]. Results on equivalence and containment of the languages of constraint automata [25] and failure based equivalences [43] provide opportunities for analysis and optimization of Reo connectors.

A constraint automaton essentially captures all behavior alternatives of a Reo connector. Therefore, it can be used to generate a state-machine implementing the behavior of Reo connectors, in a chosen target language, such as Java or C, as explained in the next section.

Variants of the constraint automata model have been devised to capture time-sensitive behavior [11,53,54], probabilistic behavior [20], stochastic behavior [26], context sensitive behavior [29,34,41], fairness [30,42], resource sensitivity [66], and the QoS aspects [12,13,67,70,71] of Reo connectors and composite systems.

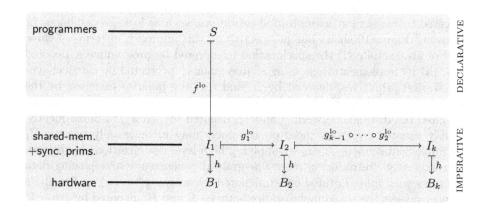

Fig. 19. From declarative specifications to imperative implementations

6   Compilation

By now, we may have convinced our readers that both (1) exogenous specification of multi-party interaction protocols (regardless of the language in which they are implemented), and (2) high-level languages that support specification of such protocols as composition of primitive interactions (as opposed to in terms of low level communication actions) offer clear software engineering advantages (e.g., programmability, maintainability, reusability, verifiability, etc.). Reo serves as a prime example of a high-level language, based on an exogenous interaction-centric model of concurrency, that demonstrates the viability of raising the level of abstraction in specification of concurrency protocols to where these software engineering advantages can indeed materialize. It seems far less obvious, however, that protocol specifications expressed in such high-level languages can be compiled into efficient and scalable binaries.

In this section, we intend to persuade the reader that in time, sufficiently smart compilers for high-level protocol languages can produce binaries with better performance than binaries produced by compilers for contemporary general-purpose languages that offer the lower-level constructs of traditional models of concurrency. At the core of our argument lies the observation that compilers for such high-level protocol languages can optimize concurrent programs in novel ways inconceivable for compilers that receive lower-level constructs of traditional models of concurrency as their input. In making this argument, first, we need to understand the limitations of compiling protocols coded in lower-level action-based languages as Java and C.

Essentially, to write a concurrent program, concurrent programmers cross a distance between a declarative specification of its protocols and processes (or threads, components, services, actors, agents, etc.), which abstractly defines *what* must happen, and its imperative implementation, which concretely defines *how* things happen. Today, the processes in such imperative implementations typically interact with each other through actions that manipulate shared-memory protected by classical synchronization primitives, such as locks, semaphores, or monitors.[4] Figure 19 shows our perspective on this approach in terms of three levels of abstraction: (i) the specification interpreted by programmers, denoted by S, (ii) its implementations using shared-memory protected by classical synchronization primitives, denoted by I_i, and (iii) the binaries executed by the hardware, denoted by B_i. In writing their concurrent program, programmers first cross the distance between S and I_1, denoted by arrow f^{lo}. Subsequently, possibly assisted by tools, these programmers may incrementally improve I_1 into implementations $I_2 \ldots I_k$ by applying high-level optimizations to the program logic (e.g., introducing more fine-grained concurrency or replacing data structures with more optimal ones), denoted by arrows $g_1^{lo} \ldots g_{k-1}^{lo}$. Finally, a compiler crosses the remaining distance between I_i and B_i, denoted by arrow h.

Figure 19 provides another perspective on the previously identified difficulties with low-level action-based concurrency. Essentially, these difficulties arise from the conceptually long distance between the levels of abstraction of S and I_1, effectively measured by comparing the textual length of specification S with the number of lines of code of its implementation I_1. Intuitively, as this ratio gets smaller, the distance between S and I_1 grows longer, and consequently, the amount of intellectual work that programmers need to perform becomes larger. In practice, it typically requires a substantial effort and significant ingenuity from programmers to define f^{lo} (i.e., to write their concurrent program with action-based concurrency) and to establish $f^{lo}(S) \sqsubseteq S$ in terms of the low-level code that $f^{lo}(S)$ consists of (i.e., to establish that f^{lo} faithfully implements S). Hamberg and Vaandrager, for instance, discuss these issues in more detail, from the perspective of teaching concurrency through model checking [39].

Additionally, Fig. 19 also shows that facing a traditional low-level action-based model of concurrency, forces programmers to take responsibility for defining, selecting, and applying every g_i^{lo} (i.e., defining, selecting, and applying optimizations) and, again, for establishing $(g_{k-1}^{lo} \circ \cdots \circ g_1^{lo} \circ f^{lo})(S) \sqsubseteq S$. Ideally, of course, a compiler instead of programmers should perform every g_i^{lo}. But although sixty years of research in compiler technology has resulted in a battery of many important low-level optimization techniques, current compilers typically cannot apply higher-level, "intention-preserving" optimizations to the program logic. For instance, automatic parallelization of general algorithms and data structures remains an open problem to this day [18].

To further illustrate this point, Fig. 20 shows the problem that a low-level compiler faces in applying such high-level "intention-preserving" optimizations. For such a compiler to decide which optimizations it can—and should—apply

---

[4] Of course, in a distributed memory setting, the concurrency primitives are different, but message passing communication primitives used in such settings still constitute an action-based model of concurrency, for which our subsequent argument still holds.

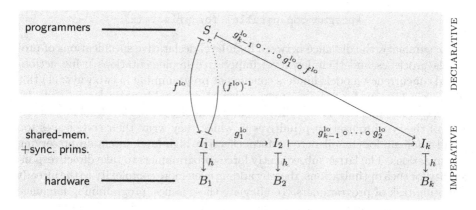

Fig. 20. Irresurrectability of declarative specifications

to which parts of implementation I_1, it essentially needs to reconstruct specification S. Only then, when the compiler knows the *itentions* that programmers had when they wrote I_1, can it decide which portions of the code admit which intention-preserving optimization. In other words, before the compiler can optimize anything, it first needs to apply the *inverse* of f^{lo} to $f^{\text{lo}}(S)$ to resurrect the lost *what*, S. Generally, however, the compiler cannot do this: in going from a declarative specification to an imperative implementation, certain information gets irretrievably lost or becomes practically impossible to extract from the resulting code. Consider, for instance, the following C code:

```
int x;
for (int i = 0; i < 10; i++) {
  x = rand();
  a[i] = some_function(x); // without side effects
}
```

If we intended *just* to assign the output of some_function to every a[i], for random inputs x, a compiler can parallelize the loop. However, if we *additionally* intended the resulting array to have the same content in executions with the same random seed (e.g., to reproduce bugs), a compiler cannot parallelize the loop: in that case, the order of generating random numbers matters. Just from this code, thus, neither a compiler *nor a human* can judiciously decide about loop parallelization; to make that decision, one needs more information.

For more complex programs, as the distance between specifications and their implementations becomes longer, the distance between those implementations and their binaries becomes relatively shorter, leaving less room for a compiler to perform significant high-level "intention preserving" optimizations. Incidentally, the annotations used in some parallelization frameworks (e.g., OpenMP) explicitly preserve information that otherwise gets lost in translation, which the compiler subsequently leverages to produce more optimized binaries. For instance, with OpenMP, a programmer can annotate the loop in the previous C code with the following pragma to inform the compiler that it may parallelize:

```
#pragma omp parallel for private(x)
```

In summary, the distance between high-level declarative specifications of protocols/processes and their low-level imperative implementations using action-based concurrency models hinders concurrent programming in two ways: (1) this distance is too long for average programmers to reasonably write correct code, let alone, correct code that is also efficient, and scalable; (2) the low level of abstraction of the synchronization primitives in which they write their code leaves too small a domain for compilers to perform effective, high-level, intention-preserving optimizations. The latter subsequently forces programmers to take direct responsibility for such optimizations, thereby adding even more complexity to the already daunting task of programmers. To alleviate these issues, programming language designers should provide programmers new, declarative, high-level interaction-based abstractions for implementing parallel programs. In the previous sections, we already argued that languages that offer such constructs, as Reo, can alleviate the first issue. Here, Figs. 19 and 20 give us the right context to argue that such languages also alleviate the second issue.

Figure 21 shows our proposed approach, where $M_1 \ldots M_k$ denote implementations of S in a special, declarative protocol language (imagine Reo). The shorter distance between S and M_1 simplifies programmers' task of writing their parallel program, denoted by arrow f^{hi}, essentially because those programmers need to concern themselves with fewer details (e.g., seemingly nondeterministic scheduling). Moreover, as programmers express their protocols at a high level of abstraction, declaratively, more information about their intentions remains available in the resulting code. A compiler can subsequently leverage this information to generate more optimized binaries. As such, this compiler relieves programmers from the responsibility of manually implementing, and establishing the correctness of, not only low-level optimizations (as current compilers already do) but also high-level intention-preserving optimizations: an application programmer now needs to work out only f^{hi}, after which the compiler takes care of selecting

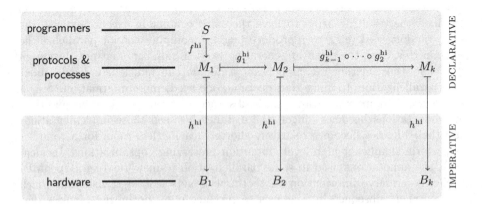

Fig. 21. From declarative specifications to declarative implementations

and applying every applicable g_i^{hi} defined by its designer. This designer, instead of application programmers, should prove the correctness and effectiveness of every g_i^{hi}, and establishing those properties remains a one-shot activity (cf. ad-hoc reasoning about every manually optimized low-level concurrent program). Moreover, because M_i and M_{i+1} reside at a higher level of abstraction than I_i and I_{i+1} do, proving the correctness and effectiveness of g_i^{hi} typically becomes simpler, clearer, and more mathematically elegant than reasoning about the low-level code manipulated by g_i^{lo}. Shortly, we give concrete examples for this claim.

Thus, by offering a new level of interaction-based abstraction to programmers, our proposed approach alleviates the software engineering difficulties of expressing implementations using action-based models of concurrency, by shortening the distance between specifications and their implementations, which in turn makes it more reasonable for programmers to perform the intellectual work required to cross this distance. Perhaps surprisingly, a shorter distance between specification and implementation has another significant advantage: it makes the distance between implementations and binaries long enough for compilers to perform also high-level intention-preserving optimizations, which also ameliorates the difficulties of developing implementations with good scalability and performance. In time, binaries generated by sufficiently smart compilers for high-level protocol languages should outperform binaries of low-level code hand-written by average programmers. In posing this thesis, we feel encouraged by the observation that although the distance between an implementation of a typical sequential program expressed in a conventional imperative languages (e.g., Java or C) and an optimized version of its binary code is also huge, the compiler construction community has still succeeded to develop effective tools for crossing this distance, demanding little or no intellectual effort from programmers. Essentially, we propose to extend that work to high-level protocol languages for concurrent programming. By now, the concrete preliminary results of experiments with our Reo compiler support our thesis and exemplify its feasibility.

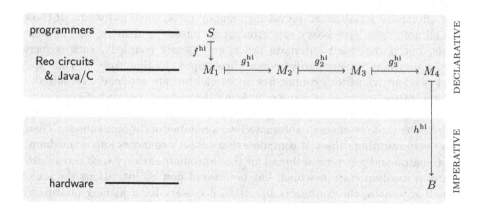

Fig. 22. From declarative specifications to imperative implementations via CAS

Figure 22 shows the instantiation of Fig. 21 in the context of our Reo compiler, which is based on Reo's constraint automaton semantics, presented in Sect. 5. In this instantiation, the programmers' task f^{hi} consists of (i) translating the processes in specification S into Java or C code and (ii) translating the protocol in S into a Reo connector; together, this code and the Reo connector constitute M_1. Our compiler subsequently maps every node and every channel in the Reo connector to its corresponding constraint automaton. This yields a set of "small" automata that collectively represent the connector's semantics. The compiler then translates this set of small automata into Java/C and merges the code so generated with the Java/C code for the processes. An external compiler for Java/C subsequently translates the full code base into a binary.

Our Reo compiler currently applies three high-level optimizations g_i^{hi}.

- g_1^{hi}—*Improving latency* [46]

  The most straight-forward translation of a parallel composition of small automata, which collectively model a connector's semantics, into Java/C works by generating a distinct thread for each of those automata. In this approach, every such thread executes a small state machine for its corresponding automaton, firing transitions as it reaches consensus with the other threads about their collective behavior. The distributed consensus algorithm necessary for achieving such multiparty synchronization, however, costs too much in terms of resources at run-time, which causes transitions to have high firing latency.

  Optimization g_1^{hi} aims at reducing firing latency: instead of translating a parallel composition of small automata to as many threads, g_1^{hi} first computes a single "big" automaton for that composition, similar to parallel expansion in process calculi, and generates only one thread for that automaton. This single thread executes a big state machine, free of other threads to synchronize its behavior with.

- g_2^{hi}—*Improving throughput* [45, 47, 48, 51]

  Although g_1^{hi} reduces firing latency, it does so at the cost of reduced firing throughput: by computing one big automaton out of multiple small automata, g_1^{hi} effectively serializes all parallelism among those small automata. If those small automata have heavy synchronization interdependencies, this is desirable, but if the small automata are more "loosely coupled", such sequentialization may unnecessarily reduce throughput. In that case, at run-time, independent transitions cannot fire in parallel but are artificially serialized.

  Optimization g_2^{hi} aims at improving firing throughput: instead of computing one big automaton for a parallel composition of small automata, it carefully *partitions* that set of small automata into a number of disjoint subsets. Then, for every resulting subset, it composes that subset's elements into a "medium" automaton and generates a thread for that automaton. Every such thread executes a medium state machine, but because of how g_2^{hi} partitions the set of small automata, the consensus algorithm necessary for achieving multiparty synchronization among those threads costs only little in terms of resources at run-time. Consequently, g_2^{hi} balances low latency (i.e., sequentiality) with high throughput (i.e., parallelism).

– g_3^{hi}—*Improving scalability* [49]

Before a thread can fire a transition, it must check the synchronization constraint and the data constraint of that transition. To check the synchronization constraint, a thread inspects all relevant interface nodes for a pending I/O-operation; if at least one of those nodes has no such operation, the transition cannot fire. To check the data constraint, a thread calls a constraint solver to find a solution for that constraint; if no solution exists, the transition cannot fire. Whenever a transition does fire, its executing thread effectively effectuates an interaction among processes. Typically, as the number of processes increases, the number of transitions per medium automaton also increases. Because the thread for such a medium automaton needs to check all its transitions for enabledness (in the worst case), firing a transition requires increasingly more resources as the number of processes increases. This suggests poor scalability.

Optimization g_3^{hi} aims at improving scalability: instead of directly generating a thread for a medium CA, it first *merges* certain distinguished transitions of that automaton into a single transition in a semantics-preserving way. Subsequently, it translates the resulting automaton, with merged transitions, into a thread. This thread executes in the same way as before, but it can check merged transitions for enabledness with a single operation, instead of with one operation per transition. Crucially, to facilitate such combined checks, g_3^{hi} injects optimized data structures for pending I/O-operations on nodes in the generated code. Because not all transitions can be merged in a semantics-preserving way, g_3^{hi} performs static analysis on the transitions of an automaton to determine the extent to which it can introduce such optimized data structures.

We have proved the correctness of the above high-level optimizations in terms of constraint automata (see their respective references, above).

The Java bytecode obtained using our compiler (and an external Java compiler afterward) runs on a JVM as any other Java program. With C, as an extra low-level optimization, we use a framework that allows instructions to be scheduled directly to cores instead of indirectly via the operating system's scheduler [49].

Compilers for low-level languages seem incapable of performing similar optimizations as those in Fig. 22 (i.e., automatic sequentialization, automatic parallelization, and automatic optimization of data structures). As practical evidence, if those compilers would be capable of this, we would have relied on those capabilities of theirs instead of developing optimizations ourselves. More philosophically, we believe that low-level compilers will never be capable of optimizing in this way, simply because they do not have enough information about programmers' intentions (see, e.g., the example of assigning random numbers to array elements, mentioned earlier in this section). Constraint automata, in contrast, retain enough such information to allow more effective high-level optimizations. At the same time, it may be difficult for average programmers to detect when and how optimizations similar to the ones in Fig. 22 may and should be applied manually; compilers for high-level protocol languages alleviate this burden.

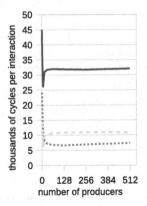

Fig. 23. Earlier performance results [50] (Color figure online)

For some protocols, the high-level intention-preserving optimizations in Fig. 22 already allow our compiler to generate code that can compete with code written by a competent programmer [50]. Figure 23 shows one of our most promising achievements so far. It shows the performance of three implementations of a k-producers-single-consumers protocol, for $k \in \{2^i \mid 2 \leq i \leq 9\}$: one naive hand-written implementation in C (blue, solid line), one hand-crafted optimized implementation in C (yellow, dashed line), and one implementation expressed in Reo and compiled via CAs into C (red, dotted line). In every round of this protocol, every producer sends one datum to the consumer. Once the consumer has received a datum from every producer, in any order, it sends an acknowledgment to the producers, thereby signaling that the consumer is ready for the next round. To measure just the performance of the protocol, we did not give the producers and the consumers real computational tasks (i.e., the producers sent only dummy data). This example shows that already our current compilation technology is capable of generating code that can compete with—and in this case even outperform—carefully hand-crafted code. Surely, our technology is not yet mature enough to always achieve such positive results. Nevertheless, this example offers preliminary evidence that programming protocols among threads using high-level, interaction-based constructs and abstractions can result in equally good—or better—performance as compared to hand-crafted code using conventional low-level, action-based models of concurrency.

The obvious superficial "performance comparison" depicted in Fig. 23 may say as much about the effectiveness of our optimization techniques, as it does about the competency of the C programmer who produced the hand-crafted version of the protocol code of this application. However, below this surface, lies a more crucial fundamental point that is independent of the competency of any individual programmer, or the precise factor by which our optimization techniques potentially can or currently do outperform hand-crafted code that a programmer can (even hypothetically) produce. Crucial to this benchmark is the fact that the task assigned to the programmer restricted him to use concurrency constructs available in contemporary programming languages, such as Java or C (in this case p-threads). On the other hand, our Reo compiler bypasses this level of abstraction (and the coarser-grained, OS-level scheduling inefficiencies that it entails) and generates code using finer-grained constructs *below* the OS-level and the concurrency constructs that it supports. From this perspective, comparing the performance of the two versions of the code is even unfair, because the statement of his task assignment prevents the programmer from using lower-level constructs to directly hand-craft code similar to (or even better than) what our Reo compiler produces. But precisely this *unfairness* constitutes the crux of our argument in this section.

There are two conceivable ways to make such a comparison fair, i.e., produce code using constructs that are "fairly comparable" to the constructs that our Reo compiler uses to produces its code: (1) allow the programmer to directly code below the level of p-threads and OS; or (2) develop tools that take p-threads level code written by a programmer and produce more optimized code.

Option 1, i.e., removing the artificial barrier of programming at the level of p-threads, is certainly possible. However, programming below p-threads and OS-level sharply raises the level of expertise required by a programmer to code directly at such a low level, and dramatically increases the size and the complexity of the resulting code. Higher competency requirements and increased size and complexity of code, in turn, sharply reduce the number of competent individuals who qualify to perform such programming assignments, and dramatically lower the likelihood of success of those who undertake such daunting tasks. Besides, applications that directly use constructs below p-threads or OS abstractions become highly brittle and non-portable, as they rely on constructs that most likely do not exist verbatim on other platforms, or even on a future upgrade of their original platforms.

Option 2 requires developing tools that can reconstruct the intentions behind the p-threads constructs used to encode a protocol (fragment). As a concrete example, a single transition in a constraint automaton may declare a complex multi-party synchronization. By the time that a programmer expresses this intention in terms of semaphores, locks, guards, communication primitives, and data structures, and intersperses its resulting code with other fragments of code that are not directly related to this specific (multi-party synchronization) intention, it becomes extremely difficult, if not theoretically impossible, for any tool to reconstruct the original intention. Not having this information prevents a tool from performing intention-preserving optimizations to generate lower-level code that can more efficiently implement an application-specific multi-party synchronization.

Offering programmers higher-level protocol specification languages, such as Reo, which directly capture and retain more of the intentions behind protocol fragments, seems like a very promising alternative. Our work on Reo and our preliminary experiments with our Reo compiler suggest this approach is a viable alternative.

7   Concluding Remarks

Action and interaction offer dual perspectives on concurrency. Execution of actions involving shared resources by independent processes that run concurrently, induces pairings of those actions, along with an ordering of those pairs, that we commonly refer to as interaction. Dually, interaction can be seen as an external relation that constrains the pairings of the actions of its engaged processes and their ordering. The traditional action-centric models of concurrency generally make interaction protocols intangible by-products, implied by nebulous specifications scattered throughout the bodies of their engaged processes. Specification, manipulation, and analysis of such protocols are possible only indirectly, through specification, manipulation, and analysis of those scattered actions, which is often made even more difficult by the entanglement of the data-dependent control flow that surrounds those actions. The most challenging aspect of a concurrent system is *what* its interaction protocol does. In contrast to the *how* which an imperative programming language specifies, declarative programming, e.g., in functional and constraint languages, makes it easier to directly specify, manipulate, and analyze the properties of *what* a program does, because *what* is precisely what they express. Analogously, in an interaction-centric model of concurrency, interaction protocols become tangible first-class constructs that exist explicitly as (declarative) constraints outside and independent of the processes that they engage. Specification of interaction protocols as declarative constraints makes them easier to manipulate and analyze directly, and makes it possible to compose interaction protocols and reuse them.

The coordination language Reo is a premier example of a formalism that embodies an interaction-centric model of concurrency. We used examples of Reo connectors to illustrate the flavor of programming pure interaction protocols. Expressed as explicit declarative constraints, protocols espouse exogenous coordination. Our examples showed the utility of exogenous coordination in yielding loosely-coupled flexible systems whose components and protocols can be easily scaled or modified, even at run time.

In addition to software engineering advantages, high-level languages to specify multi-party exogenous interaction protocols, such as Reo, have advantages with respect to performance as well: as evidenced by our Reo compiler, compilers for such high-level languages can perform optimizations that compilers for lower-level languages cannot apply.

References

1. CADP home page. http://www.inrialpes.fr/vasy/cadp/
2. Extensible Coordination Tools home page. http://reo.project.cwi.nl/cgi-bin/trac.cgi/reo/wiki/Tools
3. mCRL2 home page. http://www.mcrl2.org
4. Reo home page. http://reo.project.cwi.nl
5. Vereofy home page. http://www.vereofy.de/

6. Agha, G.: Actors: A Model of Concurrent Computation in Distributed Systems. MIT Press, Cambridge (1986)
7. Arbab, F.: Reo: a channel-based coordination model for component composition. Math. Struct. Comput. Sci. **14**(3), 329–366 (2004)
8. Arbab, F.: Abstract behavior types: a foundation model for components and their composition. Sci. Comput. Program. **55**(1–3), 3–52 (2005)
9. Arbab, F.: Puff, the magic protocol. In: Agha, G., Danvy, O., Meseguer, J. (eds.) Formal Modeling: Actors, Open Systems, Biological Systems. LNCS, vol. 7000, pp. 169–206. Springer, Heidelberg (2011)
10. Arbab, F., Aştefănoaei, L., de Boer, F.S., Dastani, M., Meyer, J.-J., Tinnermeier, N.: Reo connectors as coordination artifacts in 2APL systems. In: Bui, T.D., Ho, T.V., Ha, Q.T. (eds.) PRIMA 2008. LNCS (LNAI), vol. 5357, pp. 42–53. Springer, Heidelberg (2008)
11. Arbab, F., Baier, C., de Boer, F.S., Rutten, J.J.M.M.: Models and temporal logical specifications for timed component connectors. Softw. Syst. Model. **6**(1), 59–82 (2007)
12. Arbab, F., Chothia, T., Meng, S., Moon, Y.-J.: Component connectors with QoS guarantees. In: Murphy, A.L., Vitek, J. (eds.) COORDINATION 2007. LNCS, vol. 4467, pp. 286–304. Springer, Heidelberg (2007)
13. Arbab, F., Chothia, T., van der Mei, R., Meng, S., Moon, Y.-J., Verhoef, C.: From coordination to stochastic models of QoS. In: Field and Vasconcelos [35], pp. 268–287
14. Arbab, F., Kokash, N., Meng, S.: Towards using Reo for compliance-aware business process modeling. In: Margaria, T., Steffen, B. (eds.) ISoLA 2008. CCIS, vol. 17, pp. 108–123. Springer, Heidelberg (2008)
15. Arbab, F., Mavaddat, F.: Coordination through channel composition. In: Arbab, F., Talcott, C. (eds.) COORDINATION 2002. LNCS, vol. 2315, pp. 22–39. Springer, Heidelberg (2002)
16. Arbab, F., Meng, S.: Synthesis of connectors from scenario-based interaction specifications. In: Chaudron, M.R.V., Ren, X.-M., Reussner, R. (eds.) CBSE 2008. LNCS, vol. 5282, pp. 114–129. Springer, Heidelberg (2008)
17. Arbab, F., Meng, S., Moon, Y.-J., Kwiatkowska, M.Z., Qu, H.: Reo2MC: a tool chain for performance analysis of coordination models. In: van Vliet, H., Issarny, V. (eds.) ESEC/SIGSOFT FSE, pp. 287–288. ACM, New York (2009)
18. Arvind, D.A., Pingali, K., Chiou, D., Sendag, R., Yi, J.: Programming multicores: do applications programmers need to write explicitly parallel programs? IEEE Micro **30**(3), 19–33 (2010)
19. Baeten, J.C.M., Weijland, W.P.: Process Algebra. Cambridge University Press, Cambridge (1990)
20. Baier, C.: Probabilistic models for Reo connector circuits. J. Univers. Comput. Sci. **11**(10), 1718–1748 (2005)
21. Baier, C., Blechmann, T., Klein, J., Klüppelholz, S.: Formal verification for components and connectors. In: de Boer, F.S., Bonsangue, M.M., Madelaine, E. (eds.) FMCO 2008. LNCS, vol. 5751, pp. 82–101. Springer, Heidelberg (2009)
22. Baier, C., Blechmann, T., Klein, J., Klüppelholz, S.: A uniform framework for modeling and verifying components and connectors. In: Field and Vasconcelos [35], pp. 247–267
23. Baier, C., Blechmann, T., Klein, J., Klüppelholz, S., Leister, W.: Design and verification of systems with exogenous coordination using vereofy. In: Margaria, T., Steffen, B. (eds.) ISoLA 2010, Part II. LNCS, vol. 6416, pp. 97–111. Springer, Heidelberg (2010)

24. Baier, C., Klein, J., Klüppelholz, S.: Modeling and verification of components and connectors. In: Bernardo, M., Issarny, V. (eds.) SFM 2011. LNCS, vol. 6659, pp. 114–147. Springer, Heidelberg (2011)
25. Baier, C., Sirjani, M., Arbab, F., Rutten, J.J.M.M.: Modeling component connectors in Reo by constraint automata. Sci. Comput. Program. **61**(2), 75–113 (2006)
26. Baier, C., Wolf, V.: Stochastic reasoning about channel-based component connectors. In: Ciancarini, P., Wiklicky, H. (eds.) COORDINATION 2006. LNCS, vol. 4038, pp. 1–15. Springer, Heidelberg (2006)
27. Bergstra, J.A., Klop, J.W.: Process algebra for synchronous communication. Inf. Control **60**, 109–137 (1984)
28. Blechmann, T., Baier, C.: Checking equivalence for Reo networks. Electr. Notes Theor. Comput. Sci **215**, 209–226 (2008)
29. Bonsangue, M.M., Clarke, D., Silva, A.: Automata for context-dependent connectors. In: Field and Vasconcelos [35], pp. 184–203
30. Bonsangue, M.M., Izadi, M.: Automata based model checking for reo connectors. In: Arbab, F., Sirjani, M. (eds.) FSEN 2009. LNCS, vol. 5961, pp. 260–275. Springer, Heidelberg (2010)
31. Clarke, D., Costa, D., Arbab, F.: Modelling coordination in biological systems. In: Margaria, T., Steffen, B. (eds.) ISoLA 2004. LNCS, vol. 4313, pp. 9–25. Springer, Heidelberg (2006)
32. Clarke, D., Costa, D., Arbab, F.: Connector colouring I: Synchronisation and context dependency. Sci. Comput. Program. **66**(3), 205–225 (2007)
33. Clarke, D., Proença, J., Lazovik, A., Arbab, F.: Channel-based coordination via constraint satisfaction. Sci. Comput. Program. **76**(8), 681–710 (2011)
34. Costa, D.: Formal models for context dependent connectors for distributed software components and services. Ph.D. thesis, Vrije Universiteit Amsterdam (2010). http://dare.ubvu.vu.nl//handle/1871/16380
35. Field, J., Vasconcelos, V.T. (eds.): COORDINATION 2009. LNCS, vol. 5521. Springer, Heidelberg (2009)
36. Fokkink, W.: Introduction to Process Algebra. Texts in Theoretical Computer Science, An EATCS Series. Springer, Berlin (1999)
37. Grabe, I., Jaghoori, M.M., Aichernig, B.K., Baier, C., Blechmann, T., de Boer, F.S., Griesmayer, A., Johnsen, E.B., Klein, J., Klüppelholz, S., Kyas, M., Leister, W., Schlatte, R., Stam, A., Steffen, M., Tschirner, S., Xuedong, L., Yi, W.: Credo methodology: modeling and analyzing a peer-to-peer system in credo. Electr. Notes. Theor. Comput. Sci. **266**, 33–48 (2010)
38. Groote, J.F., Mathijssen, A., Reniers, M.A., Usenko, Y.S., van Weerdenburg, M.: The formal specification language mCRL2. In: Brinksma, E., Harel, D., Mader, A., Stevens, P., Wieringa, R. (eds.) MMOSS. Dagstuhl Seminar Proceedings, vol. 06351. Internationales Begegnungs- und Forschungszentrum fuer Informatik (IBFI), Schloss Dagstuhl, Germany (2006)
39. Hamberg, R., Vaandrager, F.: Using model checkers in an introductory course on operating systems. Oper. Syst. Rev. **42**(6), 101–111 (2008)
40. Hoare, C.A.R.: Communicating Sequential Processes. Prentice-Hall, Upper Saddle River (1985)
41. Izadi, M., Bonsangue, M.M., Clarke, D.: Modeling component connectors: synchronisation and context-dependency. In: Cerone, A., Gruner, S. (eds.) SEFM, pp. 303–312. IEEE Computer Society, Los Alamitos (2008)
42. Izadi, M., Bonsangue, M.M., Clarke, D.: Büchi automata for modeling component connectors. Softw. Syst. Model. **10**(2), 183–200 (2011)

43. Izadi, M., Movaghar, A.: Failure-based equivalence of constraint automata. Int. J. Comput. Math. **87**(11), 2426–2443 (2010)

44. Jongmans, S.-S., Arbab, F.: Overview of thirty semantic formalisms for Reo. Sci. Ann. Comput. Sci. **22**(1), 201–251 (2012)

45. Jongmans, S.-S.T.Q., Arbab, F.: Global consensus through local synchronization. In: Canal, C., Villari, M. (eds.) ESOCC 2013. CCIS, vol. 393, pp. 174–188. Springer, Heidelberg (2013)

46. Jongmans, S.-S., Arbab, F.: Modularizing and specifying protocols among threads.In: Proceedings of PLACES 2012. EPTCS, vol. 109, pp. 34–45. CoRR (2013)

47. Jongmans, S.-S., Arbab, F.: Toward sequentializing overparallelized protocol code. In: Proceedings of ICE 2014. EPTCS, vol. 166, pp. 38–44. CoRR (2014)

48. Jongmans, S.-S., Arbab, F.: Can high throughput atone for high latency in compiler-generated protocol code? In: Proceedings of FSEN 2015. Springer (in press)

49. Jongmans, S.-S.T.Q., Halle, S., Arbab, F.: Automata-based optimization of inter-action protocols for scalable multicore platforms. In: Kühn, E., Pugliese, R. (eds.) COORDINATION 2014. LNCS, vol. 8459, pp. 65–82. Springer, Heidelberg (2014)

50. Jongmans, S.-S., Halle, S., Arbab, F.: Reo: a dataflow inspired language for mul-ticore.In: Proceedings of DFM 2013, pp. 42–50. IEEE (2014)

51. Jongmans, S.-S., Santini, F., Arbab, F.: Partially-distributed coordination with Reo.In: Proceedings of PDP 2014, pp. 697–706. IEEE (2014)

52. Jongmans, S.-S.T.Q., Krause, C., Arbab, F.: Encoding context-sensitivity in reo into non-context-sensitive semantic models. In: De Meuter, W., Roman, G.-C. (eds.) COORDINATION 2011. LNCS, vol. 6721, pp. 31–48. Springer, Heidelberg (2011)

53. Kemper, S.: SAT-based verification for timed component connectors. Electr. Notes Theor. Comput. Sci. **255**, 103–118 (2009)

54. Kemper, S.: Compositional construction of real-time dataflow networks. In: Clarke, D., Agha, G. (eds.) COORDINATION 2010. LNCS, vol. 6116, pp. 92–106. Springer, Heidelberg (2010)

55. Klein, J., Klüppelholz, S., Stam, A., Baier, C.: Hierarchical modeling and formal verification. An industrial case study using reo and vereofy. In: Salaün, G., Schätz, B. (eds.) FMICS 2011. LNCS, vol. 6959, pp. 228–243. Springer, Heidelberg (2011)

56. Klüppelholz, S., Baier, C.: Symbolic model checking for channel-based component connectors. Electr. Notes Theor. Comput. Sci **175**(2), 19–37 (2007)

57. Koehler, C., Lazovik, A., Arbab, F.: ReoService: coordination modeling tool. In: Krämer et al. [64], pp. 625–626

58. Kokash, N., Krause, C., de Vink, E.P.: Data-aware design and verification of service compositions with Reo and mCRL2. In: SAC 2010: Proceedings of the 2010 ACM Symposium on Applied Computing, pp. 2406–2413. ACM, New York (2010)

59. Kokash, N., Arbab, F.: Formal behavioral modeling and compliance analysis for service-oriented systems. In: de Boer, F.S., Bonsangue, M.M., Madelaine, E. (eds.) FMCO 2008. LNCS, vol. 5751, pp. 21–41. Springer, Heidelberg (2009)

60. Kokash, N., Arbab, F.: Applying Reo to service coordination in long-running busi-ness transactions. In: Shin, S.Y., Ossowski, S. (eds.) SAC, pp. 1381–1382. ACM, New York (2009)

61. Kokash, N., Arbab, F., Changizi, B., Makhnist, L.: Input-output conformance testing for channel-based service connectors. In: Aceto, L., Mousavi, M.R. (eds.) PACO. EPTCS, vol. 60, pp. 19–35 (2011)

62. Kokash, N., Krause, C., de Vink, E.P.: Verification of context-dependent channel-based service models. In: de Boer, F.S., Bonsangue, M.M., Hallerstede, S., Leuschel, M. (eds.) FMCO 2009. LNCS, vol. 6286, pp. 21–40. Springer, Heidelberg (2010)
63. Kokash, N., Krause, C., de Vink, E.P.: Time and data-aware analysis of graphical service models in Reo. In: Fiadeiro, J.L., Gnesi, S., Maggiolo-Schettini, A. (eds.) SEFM, pp. 125–134. IEEE Computer Society (2010)
64. Krämer, B.J., Lin, K.-J., Narasimhan, P. (eds.): ICSOC 2007. LNCS, vol. 4749. Springer, Heidelberg (2007)
65. Lazovik, A., Arbab, F.: Using Reo for service coordination.In: Krämer et al. [64], pp. 398–403
66. Meng, S., Arbab, F.: On resource-sensitive timed component connectors. In: Bonsangue, M.M., Johnsen, E.B. (eds.) FMOODS 2007. LNCS, vol. 4468, pp. 301–316. Springer, Heidelberg (2007)
67. Meng, S., Arbab, F.: QoS-driven service selection and composition. In: Billington, J., Duan, Z., Koutny, M. (eds.) ACSD, pp. 160–169. IEEE (2008)
68. Milner, R. (ed.): A Calculus of Communicating Systems. LNCS, vol. 92. Springer, Heidelberg (1980)
69. Milner, R.: Elements of interaction - turing award lecture. Commun. ACM 36(1), 78–89 (1993)
70. Moon, Y.-J.: Stochastic models for quality of service of component connectors. Ph.D. thesis, Leiden University (2011)
71. Moon, Y.-J., Silva, A., Krause, C., Arbab, F.: A compositional semantics for stochastic Reo connectors. In: Mousavi, M.R., Salaün, G. (eds.) FOCLASA. EPTCS, vol. 30, pp. 93–107 (2010)
72. Sangiorgi, D., Walker, D.: PI-Calculus: A Theory of Mobile Processes. Cambridge University Press, New York (2001)
73. Schumm, D., Turetken, O., Kokash, N., Elgammal, A., Leymann, F., van den Heuvel, W.-J.: Business process compliance through reusable units of compliant processes. In: Daniel, F., Facca, F.M. (eds.) ICWE 2010. LNCS, vol. 6385, pp. 325–337. Springer, Heidelberg (2010)
74. Wegner, P.: Coordination as comstrainted interaction (extended abstract). In: Hankin, C., Ciancarini, P. (eds.) COORDINATION 1996. LNCS, vol. 1061, pp. 28–33. Springer, Heidelberg (1996)

Modeling of Architectures

Jade Alglave[✉]

Microsoft Research Cambridge and University College London, London, UK
j.alglave@ucl.ac.uk, jaalglav@microsoft.com

Abstract. Concurrent programming is known to be quite hard. It is made even harder by the fact that, very often, the execution models of the machines we run our software on are not precisely defined.

This document is a tutorial on the herd tool and the `cat` language, in which one can define consistency models.

M. Bernardo and E.B. Johnsen (Eds.): SFM 2015, LNCS 9104, pp. 97–145, 2015.
DOI: 10.1007/978-3-319-18941-3_3

1   Why Herd Cats Anyway?

Concurrent programming is known to be quite hard.

Look at this pink pony:

It's been computed on an Intel x86 CPU, with sequential code. It's healthy!
Now look at this one:

It's been computed on an Nvidia GPU, with concurrent code. It's all broken;
this is because multicore machines sometimes do not behave quite as we'd expect,
and sensible looking code can end up producing unexpected results.

Now, there are ways to repair the concurrent pony, look; this pony was also
computed on an Nvidia GPU, under the same conditions as the previous one:

One can repair our pony by placing special instructions in the code used
to compute it. However the semantics of such special instructions is very often
poorly documented.

In this lecture we present a language, called cat, that can help us tackle this
issue by allowing us to precisely describe the semantics of these special instructions.

More precisely, our `cat` language gives us means to describe *consistency models*, i.e. execution models of concurrent and distributed systems.

By the end of this lecture, you should have been able to build several models:

- Sequential Consistency [8] (see Sect. 3)
- Total Store Order [10] (see Sect. 3),
- a model similar in spirit to IBM Power or ARM [5] (see Sect. 4),
- a model similar in spirit to Nvidia GPUs [3] (see Sect. 5),
- a model inspired by C++ [14] (see Sect. 6).

Reading Notes. Most of the lecture is going to be interactive, using the herd tool. For this tutorial to go smoothly, I would suggest to:

- have internet access, to be able to go to the herd web interface: http://virginia. cs.ucl.ac.uk/herd-web/?book=tutorial;
- open the appendix, where I give the final models that the tutorial should help you build; these can act as guidelines along the way.

2   First Steps in Herding Cats

Here we're going to learn about the essential concepts to follow this lecture.

Litmus Tests. Litmus tests are small snippets of assembly or pseudo-assembly code that allow us to examine the behaviour of a chip (see e.g. [5]), or a model like we're doing here.

Below is our first litmus test. In this test, called MP (short for *message passing*), two processors P0 and P1 communicate via two shared memory locations x and y, both initialised to 0:

```
Bell MP
{
x = 0;
y = 0;
}
 P0          | P1          ;
 w[] x 1  | r[] r1 y  ;
 w[] y 1  | r[] r2 x  ;
exists (1:r1 = 1 /\ 1:r2 = 0)
```

On the left, the thread P0 writes 1 to memory location x, and 1 to memory location y. On the right, the thread P1 reads from y and places the result into register r1, and reads from x and places the result into register r2. The registers r1 and r2 are private to P1.

Essentially, P0 writes a message in x, then sets up a flag in y, so that when P1 sees the flag (via its read from y), it can read the message in x.

At the bottom of the test, we ask "is there an execution of this test such that register r1 contains the value 1 and register r2 contains the value 0?".

Exercise: What do you Think? Do you think such an execution is possible?

The Herd Tool. Now, let's try it out! The herd tool lets us simulate a model, and run litmus tests against that model, to determine which executions are allowed by this model. So let's go to virginia.cs.ucl.ac.uk/herd-web/?book=tutorial. In the "litmus test" box, find the file mp.litmus in the drop box. Let's select "all executions" for the display, and then click on the pink pony.

Histograms. In the "histogram" box, we see the following result:

```
Test MP Allowed
States 4
1:r1=0; 1:r2=0;
1:r1=0; 1:r2=1;
1:r1=1; 1:r2=0;
1:r1=1; 1:r2=1;
Ok
```

Note that we see the result 1:r1=1; 1:r2=0;, which we asked about in our test. Thus this result is reachable by our test, and this is why the tool says Ok; otherwise it would say No.

Executions. In the "executions" box, we see that this test can have four different executions. In all of them we see four *events*: writes to x or y, of the shape W[] x=32, which means that value 32 was written to the memory location x, and reads from x or y, of the shape R[] y=52, meaning that memory location y was read, and the value read was 52. We also see two types of *relations* over these events: po arrows and rf arrows.

Program Order. The po arrows represent the *program order*. The program order intuitively corresponds to the order in which the events occur on a thread.

For example in our test, on P0, the write of x is appears in program order before the write of y. Therefore the two corresponding events in Fig. 1 are related by po. Similarly the read of y on P1 appears in our test in program order before the read of x; thus the two corresponding events in Fig. 1 are related by po.

Note also that po is transitive. In Fig. 1, suppose there was an extra event *e* occuring on P1 with a po arrow from *d* to *e*. In this case there would also be a po arrow from *c* to *e*. That is, *e* is after *c* in the program order.

Read-from. The rf (*read-from*) arrows depict who reads from where; more precisely, for each read of a certain location, it finds a unique write of the same location and value.

Let's go through each execution one by one. In the first one, the two reads on P1 read from the initial state, which we depict with rf arrows with no sources. Our test has initialised x and y to 0, thus the read events in the first execution have the value 0. In the second execution, the read of x reads from the update of x by P0, hence has the value 1; however the read of y reads from the initial state still, hence takes the value 0. The third execution is the one our test was asking about: the read of y reads from the update by P0, whereas the read of

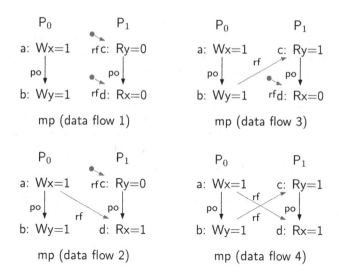

Fig. 1. Four executions of the MP test

x takes its value from the initial state, i.e. 0. In the last execution, both reads take their values from the updates by P0, i.e. they both read the value 1.

Now we're going to write a model that will forbid the result `1:r1=1; 1:r2=0;`, which we asked about in our test. It won't be a minimal model forbidding this result though (by which I mean that the model will forbid much more behaviours than this particular one), but that's okay.

To do so, we need to learn a few more concepts.

Coherence Order. In most models I know of, there's a notion of a *coherence order*. Intuitively it's a history of all writes to a given memory location x, that represents the order in which writes to x hit x. If you were sitting in memory location x, looking at writes falling down onto you, and recording their values as they come by, you would get the coherence order. In this lecture, the coherence order co is a total order over writes to a given memory location.

From-read. Using the read-from and coherence relations, we can build a relation called *from-read*:

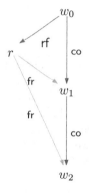

Intuitively, a from-read arrow starts from a read of a given location x, for example the read r just above, and points to all the writes that overwrite the value this read has taken. In the drawing above, the read r takes its value from the write w_0, as shown by the rf arrow between them. The write w_0 is then overwritten by the write w_1, as shown by the co arrow between them. Hence there is an fr arrow between r and w_1, as w_1 overwrites the write w_0 from which r reads. The write w_1 is then overwritten by the write w_2, as shown by the co arrow between them. Thus there is an fr arrow between r and w_2 too.

A good place to build fr is our cat file: go to the "model" box, and find the "toggle cat" button. The interface should put you in front of the cat file called tutorial.cat. If you click on "make custom cat", you should be able to edit it, and write the line above underneath the title and preamble of the cat file, i.e. underneath:

```
"I can't dance"
```

```
include "tutorial.cat"
```

In cat speak, this is how we can build the from-read fr (you can put this line in your cat file):

```
let fr = rf^-1;co
```

which means that two events a and b are related by fr if it's possible to follow an rf arrow backwards from a to some event, then a co arrow forwards from there, arriving at b. Also let's add

```
show fr
```

to the cat file, to make sure that the tool displays the new fr arrows.

Now let's run herd on MP again (don't forget to click on the pink pony!). In the third execution, observe the fr arrow between the read of x by P1 and the update of x by P0:

Let's look at this execution a bit more: the read of y by P1 reads the update by P0, whereas the read of x by P1 reads the initial state, i.e. ignores the update of x by P0. Note that this is an execution that leads to the result 1:r1=1; 1:r2=0; that we asked about in our litmus test.

Also, note that such an execution compromises the implementation of message passing idioms: intuitively, we'd like P1 to access the update of the data x after it has read the update of the flag y.

In our models, these kind of violations will occur as cycles. Notice, in the execution above, how we can follow the arrows to find a cycle: from a to b to c to d and back to a.

3   Let's Herd Two Kittens [12]

There are several models that would forbid this compromising execution. We'll build two: Sequential Consistency (SC) [8] and Total Store Order (TSO), which is the model of Sparc TSO [11] and Intel x86 [10]. In fact we won't build exactly TSO below, only a fragment of it, but you'll get to build TSO properly in an exercise later on.

Sequential Consistency is such that executions of a program are interleavings of the instructions appearing on the different threads of the program. One can show (see e.g. [1,2]) that this corresponds to an *axiomatic* model, i.e. model phrased in terms of events, program order, read-from, coherence and from-read (such as the ones we are manipulating here), where the program order and what I call the *communication relations* (i.e. the union of read-from, coherence and from-read) are compatible, i.e. there cannot be any cycle in their union. Let's first build the communication relations in cat speak (to put in your `cat` file):

```
let com = rf | co | fr
```

Here we're gathering the read-from `rf`, coherence order `co` and the from-read `fr` in a relation called `com` (for communications). In `cat` speak, the symbol | is the union of arrows, i.e. of relations over events.

Now we can build a *procedure* that will implement Sequential Consistency (to put in your `cat` file):

```
procedure sc() =
  let sc-order = (po | com)+
  acyclic sc-order
end
```

This procedure we call `sc`, and it enforces the acyclicity (through the keyword `acyclic`) of the relation `sc-order`. This relation is defined as the union of the program order `po` and the communication relation `com` that we built earlier. We need to call this procedure later on in the `cat` file for it to have an effect:

```
call sc()
```

Now let's run our MP example under that new `cat` file. Let's have a look at the histogram:

```
Test MP Allowed
States 3
1:r1=0; 1:r2=0;
1:r1=0; 1:r2=1;
1:r1=1; 1:r2=1;
No
```

Observe that there are now only 3 possible states, and the compromising result 1:r1=1; 1:r2=0; has disappeared. If you look at the executions, you'll see that we do not have the compromising one anymore. That's because there was a cycle in the union of the program order po and the communication com: from *a* to *b* to *c* to *d* and back to *a*.

Total Store Order. is such that write-read pairs (whether to the same memory location or not) on the same thread can appear to occur out of order. This may be due to processor caches and buffering (which lead to faster processors, but surprising behaviours).

Consider the following *store buffering* litmus test, where two processors P0 and P1 communicate via two shared memory locations x and y, both initialised to 0:

```
Bell SB
{
x = 0;
y = 0;
}
 P0          | P1         ;
 w[] x 1  | w[] y 1  ;
 r[] r1 y | r[] r2 x  ;
exists (0:r1 = 0 /\ 1:r2 = 0)
```

On the left, the thread P0 writes 1 to memory location x, and reads from y and places the result into register r1. Symmetrically, the thread P1 writes 1 to memory location y, and reads from x and places the result into register r2. The register r1 is private to P0, and r2 is private to P1.

At the bottom of the test, we ask the question "is there an execution of this test such that both registers contain the value 0?"

Let's comment out the call to SC because we're trying to build a different model here:

```
(* call sc() *)
```

Let's feed this test to herd; find the file sb.litmus in the litmus test drop box. After clicking on the pink pony, it says:

```
Test SB Allowed
States 4
0:r1=0; 1:r2=0;
0:r1=0; 1:r2=1;
0:r1=1; 1:r2=0;
0:r1=1; 1:r2=1;
Ok
```

Note that we see the result 0:r1=0; 1:r2=0;, which we asked about in our test. This result corresponds to the following execution:

$$P_0 \qquad\qquad P_1$$

a: Wx=1   c: Wy=1

po ↓   fr fr   ↓ po

b: Ry=0   d: Rx=0

sb

More precisely, the two reads on each thread reads from the initial state (e.g. the read on P1 reads from the initial state for x). Since the initial state is overwritten by the updates to x and y, we are exactly in the fr situation: for example the read of x on P1 is fr-before the write of x on P0.

Such a result can be explained for example by the presence of *store buffers*, one per thread: both writes (the write to x on P0 and the write to y on P1) can sit in their threads' store buffers for a while, then the reads (from y on P0 and from x on P1) can take their value from the initial state, i.e. from memory, and finally the writes can hit the memory.

This would be the case on an Intel x86 [10], or a Sparc TSO machine [11].

One can show (see e.g. [1,2,10]) that TSO corresponds to an axiomatic model (i.e. phrased in terms of events, program order, read-from, coherence and from-read), where a fragment of the program order and the communication relations are compatible, i.e. there cannot be any cycle in their union. Remember that we've built the communication relations earlier, and placed the definition in our cat file.

To model TSO, we also need to build a notion of read-froms between different threads, which we often call *external read-from* (in our cat file, next to the definition of com for example):

```
let rfe = rf & ext
```

where the primitive ext is the relation that gathers pairs of events that belong to different threads, such as the write of y by P0 and the read of y by P1.

Now, one way to model the store buffering scenarios allowed by TSO is to reorder write-read pairs (whether to the same memory location or not). This means that TSO differs from SC on write-read pairs. Thus we have to exclude all the write-read pairs from sc-order to build TSO. We can then require the acyclicity of this new relation tso-order. Let's build this as a procedure once again (to put in your cat file, just below the sc procedure):

```
procedure almost-tso() =
  let ppo = po \ W*R
  let tso-order = ppo | rfe | co | fr
  acyclic tso-order
end
```

Here we're declaring a procedure called almost-tso, in which there is a local definition of a relation called ppo (for preserved program order). We define ppo as the program order po, minus the write-read pairs: in cat speak the "setminus" operation is \; the (predefined) set of write events is W, the (predefined) set of read events is R, thus the set of all write-read pairs is W*R.

On the second line of this procedure, we require the acyclicity of the union (remember that the symbol | is the union in cat speak) of ppo and the communications com.

We need to call our almost-tso procedure in our cat file:

```
call almost-tso()
```

If you run the test now, you should find that the test has still not been forbidden. Now, most architectures provide special instructions called *fences* that prevent certain reorderings. For example, Intel x86 provides mfence, which prevents the reordering of write-read pairs allowed by TSO. Let's build ourselves such a fence; the right place to do so is in the bell file. Go to the "model" box, and find the "toggle bell" button; if you click on it, it should put you in front of tutorial.bell. Clicking on "make custom bell" should allow you to edit the bell file. Do not remove the title line "I can't dance"; you could write another title, but not remove the title entirely.

First we declare a possible *annotation* for our fence, for example 'wr (for write-read):

```
enum Fences = 'wr
```

More precisely, here we declare an enumeration of possible fence annotations under the name Fences. For now we only have one annotation, wr. This enum will also create a set Wr of all the events that bear the annotation wr.

Interlude: Enums, Tags, and Annotations. More precisely, in cat speak, one can define enumeration types, as follows. This should go into a bell file; if you want to try it out I would suggest opening a fresh bell file to not spoil the one we're currently building; don't forget to give a title to your file [16]:

```
"Hey Hey Mama"
enum Led = 'z || 'e || 'p
```

Here we're defining an enumeration type Led, which contains three *tags*: 'z, 'e and 'p. The user can then use these tags to specify that certain events can bear eponymous *annotations*; for example (again in the bell file):

```
events W[{'z}]
events R[{'e,'p}]
```

specifies that write events (which belong to the predefined set W) can bear the annotation z, whilst read events (which belong to the predefined set R) can bear the annotations e or p.

The user can then use these annotations in litmus tests, for example (if you want to try it out, find ledzep.litmus in the litmus test drop box):

```
Bell BlackDog
{
x = 0;
y = 0;
}
P0          | P1         ;
w[z] x 1  | r[p] r2 x ;
r[e] r1 y |            ;
exists(0:r1=0 /\ 1:r2=1)
```

Internally, herd has built one set for each possible tag that was declared in the enumeration type Led: the set Z gathers all events with annotation z, the set E all events with annotation e, and the set P all events with annotation p. Thus in the litmus test LedZep above, the write event on P0 will belong to the set Z, the read on P0 to the set E, and the read on P1 to the set P.

The user can manipulate these sets in the bell file; for example:

```
let ze = Z*E
let ep = E*P
let zep = ze;ep
```

defines three relations ze, ep, and zep, such that ze gathers all pairs where the left extremity belongs to Z and the right extremity belongs to E, ep gathers all pairs where the left extremity belongs to E and the right extremity belongs to P, and zep builds the sequence of a step of ze and a step of ep. We can add

```
show ze, ep, zep
```

if we want to visualise these three relations.

Back to the Write-read Fence. Then we need to say that our fence events (much like read and write events, but to represent fences) can bear the annotations that we've just defined:

```
events F[Fences]
```

More precisely, here we say that our fence events, of the shape f(...) can bear the annotation wr. Let's go ahead and modify the SB example by adding fences to it:

```
Bell SB+fwr+fwr
{
x = 0;
y = 0;
}
 P0          | P1         ;
 w[] x 1   | w[] y 1   ;
 f[wr]     | f[wr]     ;
 r[] r1 y | r[] r2 x  ;
exists (0:r1 = 0 /\ 1:r2 = 0)
```

As you can see, there is a fence instruction f(wr) between each write-read pair on P0 and P1. Thus the two corresponding events belong to the set Wr.

Now we need to build a *relation* to gather all pairs of memory events (read or write) that are separated by a fence in between them in program order; the standard library has such a primitive, it is called fencerel. We can put the following lines in our bell file for example (where F & Wr is the set of fence events that bear the annotation wr):

```
let fwr = fencerel(F & Wr)
show fwr
```

The definition of fencerel is in herd's standard library; it is as follows:

```
let fencerel(S) = (po & (_ * S)); po
```

It takes as argument a set S of events, and builds the sequence (the symbol; designates the sequence of relations in cat speak) of the relation po & (_ * S) and the program order po. Now let's look at the relation po & (_ * S) a bit more. It is built as the intersection (& is the intersection in cat speak) of the program order po, and the set of pairs (_ * S). These pairs are such that the domain (the left extremity) can be anything, whether read or write (_ means "anything" in cat speak), and the range (the right extremity) is in the set S.

So now when we define fwr as fencerel(F & Wr) up above, this means that we've built the relation gathering all pairs of events (read or write), such that there is a fence event (i.e. that belong to F) in between them in program order, and this fence event bears the annotation wr (i.e. belongs to the set Wr). This is true, for example, of the write-read pair on P0 in our example SB+fwr+fwr.

Now we can add this fwr relation to our tso definition; that is, we can update our almost-tso procedure as follows:

```
procedure almost-tso() =
  let ppo = po \ W*R
  let tso-order = ppo | fwr | rfe | co | fr
  acyclic tso-order
end
```

Note how we've added the fwr relation to the shape of cycles forbidden by our tso definition, so that now we can have no cycle that is made of ppo, fwr or com arrows.

Let's feed this test to herd and see what it says (find the file sb+fwr+fwr. litmus in the litmus test drop box):

```
Test SB+fwr+fwr Allowed
States 3
0:r1=0; 1:r2=1;
0:r1=1; 1:r2=0;
0:r1=1; 1:r2=1;
No
```

Observe that the final state we were asking about, that can be explained by the store buffer scenario outlined above, has disappeared, thanks to the fences.

Think of saving your `bell` and `cat` files, for example under the names `kittens.bell` and `kittens.cat`.

4   Let's Herd Our First Big Cat: A Tiger [6]

Today we'll learn how to build a model that is similar in spirit to IBM Power and ARM. These two models revolve around a handful of principles that we'll build one by one.

Start from fresh `bell` and `cat` files, but keep the definitions of `fr`, `rfe`, and `com`. We're going to use the `sc` procedure in a slightly different way, to `flag` the executions that do not satisfy SC; in our `kittens.cat` file we had:

```
procedure sc() =
  let sc-order = (po | com)+
  acyclic sc-order
end

[...]

call sc()
```

to forbid non-SC executions, i.e. executions with a cycle in the union of the program order `po` and the communications `com`.

Now we don't want to forbid SC executions, but just `flag` them. Thus we implement a procedure to flag non-SC executions (in our `cat` file):

```
procedure sc-flag() =
  let sc-order = (po | com)+
  flag ~acyclic sc-order as non-sc
end

call sc-flag()
```

that is we flag, with the name `non-sc`, all the executions where there is a cycle in the union of the program order `po` and the communications `com`. Note that ~ is the negation. Finally, select the "positive executions" display option.

4.1   SC per Location

The first principle is called SC PER LOCATION. It means that if you analyse your program through the prism of a sole memory location at a time, everything looks as if on SC. Formally, this also means that non-relational analyses are sound for free under consistency models that respect SC PER LOCATION [4].

Now, let's look at a bunch of litmus tests.

coWW

```
Bell coWW
{
x = 0;
}
 P0        ;
 w[] x 1   ;
 w[] x 2   ;
exists (x=1)
```

In the `coWW` test, we have only one thread P0, which does two writes of memory location x in program order. The first write in program order writes the value 1 to x; the second writes 2. We're asking at the end if it is possible to have the value 1 in x at the end, which means that the write of value 2 has hit the memory before the write of value 1.

Recall that we have defined the coherence order precisely for that purpose: describing the order in which writes to a given memory location hit that location.

Now let's run herd on our cat file and the test `coWW` (find `coww.litmus` in the litmus test drop box). We get the following histogram:

```
Test coWW Allowed
States 2
x=1;
x=2;
Ok
```

On the execution side, we get the execution that leads to the final state we've asked about in our litmus test; note that it's a non-SC execution (see the line `Flag non-sc` in the histogram above) because of its cycle in the union of po and com:

$$P_0$$

a: Wx=1

$\left(\; co \;\right]$ po

b: Wx=2

coWW

If you select the "all executions" option, you'll see two executions; one where the coherence order co follows the program order po, and the one above, where the coherence order co is in the opposite direction as po.

coRW1

```
Bell coRW1
{
x = 0;
}
 P0          ;
 r[] r1 x ;
 w[] x 1   ;
exists (0:r1=1)
```

In the coRW1 test, we have only one thread P0, which does a read and a write of memory location x in program order. The read access reads x and places the result into register r1. The write access writes the value 1 to x. We're asking at the end if it is possible to have the value 1 in r1 at the end, which means that the read takes its value from the po-later write.

Recall that we have defined the read-from precisely for that purpose: describing who reads from where; in this case if the read of x can take its value from the po-later write of value 1.

Now let's run herd on our cat file and the test coRW1 (find the file corw1. litmus in the litmus test dropbox). We get the following histogram:

```
Test coRW1 Allowed
States 2
0:r1=0;
0:r1=1;
Ok
```

On the execution side, we get the execution that leads to the final state we've asked about in our litmus test; note that it's a non-SC execution (see the Flag non-sc line just above) because of its cycle in the union of po and com:

$$P_0$$

a: Rx=1

$\left(po \right)_{rf}$

b: Wx=1

coRW1

If we select the "all executions" display option, we get two executions; one where the read takes its value from the initial state, and the one above, where the read takes its value from the po-later write of value 1.

coRW2

```
Bell coRW2
{
x = 0;
}
 P0         | P1      ;
 r[] r1 x   | w[] x 2 ;
 w[] x 1    |         ;
exists (0:r1=2 /\ x=2)
```

In the coRW2 test, we have two threads P0 and P1, communicating via the shared memory location x which is initialised to 0. P0 is the same as in the previous test coRW1, i.e. does a read and a write of memory location x in program order. The read access reads x and places the result into register r1. The write access writes the value 1 to x. P1 writes the value 2 into memory location x. We're asking at the end if it is possible to have the value 2 in r1 and in x at the end, which means that the read takes its value from the write of x on P1 (r1=2) and that the write by P1 hits the memory after the write by P0 (x=2).

Now let's run herd on our cat file and the test coRW2 (find corw2.litmus in the litmus test dropbox).

On the execution side, we get the following execution; note that it's a non-SC execution because of its cycle in the union of po and com:

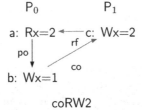

coRW2

If we select the "all executions" display option, we get six executions, the third one being the one leading to the result we're asking about in the test. In this execution, the read of x by P0 takes its value from the write by P1 (note the read-from rf arrow between them), and the write by P0 hits the memory before the write by P1 (note the coherence arrow co between them).

coWR

```
Bell coWR
{
x = 0;
}
 P0         | P1      ;
 w[] x 1    | w[] x 2 ;
 r[] r1 x   |         ;
exists (0:r1=2 /\ x=1)
```

In the coWR test, we have two threads P0 and P1, communicating via the shared memory location x which is initialised to 0. P0 writes 1 into memory location x, and reads x, placing the result into register r1. P1 writes 1 into x. At the end we're asking if it's possible to have the value 2 into r1, i.e. the read by P0 reads from the write by P1, and to have the value 1 in x, i.e. the write of P0 hits the memory after the write of P1.

Now let's run herd on our cat file and the test coWR (file cowr.litmus). On the execution side, we get the following non-SC execution:

$$P_0 \qquad\qquad P_1$$

a: Wx=1 ⟵ c: Wx=2

b: Rx=2

coWR

If we select the "all executions" option, we get six executions, the fourth one being the one leading to the result we're asking about in the test. In this execution, the read of x by P0 takes its value from the write by P1 (note the read-from rf arrow between them), and the write by P0 hits the memory after the write by P1 (note the coherence arrow co between them).

Recall that we have defined the from-read precisely for that purpose: starting from a read such as the one by P0, pointing to all the writes (such as the one by P0) that overwrite the value read (given by the write on P1). Thus we have a from-read arrow fr between the read by P0 and the po-preceding write by P0.

coRR

```
Bell coRR
{
x = 0;
}
 P0         | P1      ;
 r[] r1 x | w[] x 1 ;
 r[] r2 x |         ;
exists (0:r1=1 /\ 0:r2=0)
```

In the coRR test, we have two threads P0 and P1, communicating via the shared memory location x which is initialised to 0. P0 reads x twice, placing the results into r1 and r2; P1 writes 1 to x. At the end we're asking if it's possible for r1 to hold the value 1, i.e. after having read from P1, and for r2 to hold the value 0, i.e. after having read from the initial state.

Now let's run herd on our cat file and the test coRR (file corr.litmus).

On the execution side, we get the following execution; note that it's a non-SC execution because of its cycle in program order and communications:

$$P_0 \qquad P_1$$

a: Rx=1 ◄──── c: Wx=1
   | rf
po | ╱
   ▼   ╱ fr
b: Rx=0

coRR

If we select the "all executions" option, we get four executions, the third one being the one leading to the result we're asking about in the test. In this execution, the first read by P0 reads from the write on P1 (note the read-from arrow rf between them), and the second read by P0 reads from the initial state.

Recall that we have defined the from-read precisely for that purpose: starting from a read such as the second one by P0, pointing to all the writes (such as the one by P1) that overwrite the value read (given by the initial write of x). Thus we have a from-read arrow fr between the read by P0 and the write by P1.

Forbidding these Idioms. Let's look at all the executions that we have flagged to be non-SC: observe that they all have a similar shape, in which the program order contradicts the communication relations. More precisely, because all of our tests use one memory location only, it's the program order restricted to the same location that contradicts the communication relations.

Let's define a new notion po-loc, i.e. the program order restricted to both extremities having the same location; in cat speak (a good place to put this definition would be in the cat file, just next to the definition of com):

```
let po-loc = po & loc
```

where the primitive loc gathers all pairs of read and write events that have the same location.

Now, let's require for po-loc to not contradict our communication relations (recall we've built this notion before, into the relation com), within a procedure sc-per-location:

```
procedure sc-per-location() =
  acyclic po-loc | com
end
```

Don't forget to call the procedure sc-per-location in your cat file:

```
call sc-per-location()
```

Now let's re-run all of our tests, and observe that we do not have the executions that we've flagged anymore!

Exercise: SC per Location with Load-Load Hazard. Certain architectures, such as Sparc RMO [11] allow what is sometimes called *load-load hazard*, i.e. a situation where the coRR test that we've just seen is allowed to yield the result 0:r1=1; 0:r2=0;.

How do you think we can build a check that forbids all tests coWW, coRW1, coRW2 and coWR, but allows the test coRR?

4.2   No Thin Air

The second principle is called NO THIN AIR. Intuitively, this principle forbids scenarios where a read can take its value from a write that *depends* on this read. The word "depends" can be interpreted in many different ways; let's make that precise. Consider the following litmus test:

```
Bell LB
{
x = 0;
y = 0;
}
 PO          | P1          ;
 r[] r1 x | r[] r2 y ;
 w[] y 1  | w[] x 1  ;
exists (0:r1 = 1 /\ 1:r2 = 1)
```

In the LB test, we have two threads P0 and P1. P0 reads x and places the result into register r1, then writes 1 to y. P1 reads y and places the result into register r2, then writes 1 to x. At the end we're asking whether it is possible for both registers to contain the value 1, i.e. if the two reads can read from the po-later writes. This is perfectly well possible on ARM or Nvidia machines for example [3,5], because the read-write pairs on each thread can be reordered.

Let's run herd on this test with our current cat file (find the lb.litmus file in the litmus test drop box); we get the following histogram:

```
Test LB Allowed
States 4
0:r1=0; 1:r2=0;
0:r1=0; 1:r2=1;
0:r1=1; 1:r2=0;
0:r1=1; 1:r2=1;
Ok
```

The execution corresponding to the situation we asked about in the test is as follows; note that it's a non-SC execution because of its cycle in union of program order and communications:

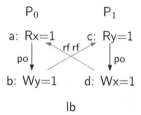

Now, we can place something between the read and the write on each thread to prevent their reordering: we can use *dependencies*, typically *address*, *data*, or *control* dependencies. Note that dependency relations always start with a read.

Interlude: Implementing Dependencies. In this lecture, we'll abstract away from actual ways of implementing dependencies. But to give an idea of what I mean, here's an example of data dependency. Consider the following variant of LB:

```
Bell LB+datas
{
x = 0;
y = 0;
}
P0              | P1              ;
r[] r1 x        | r[] r1 y        ;
xor r2 r1 r1    | xor r2 r1 r1    ;
add r3 r2 1     | add r3 r2 1     ;
w[] y r3        | w[] x r3        ;
exists (0:r1 = 1 /\ 1:r2 = 1)
```

Observe how we use operations on registers between the read and write on each thread. More precisely on P0, we read location x and place the result into register r1. Then we xor the value in r1 with itself, and place the result into register r2 (of course the result is always 0, but that's okay). Then we add 1 to the value in r2, and place the result (i.e. 1) into r3. Finally we write the value in r3 into location y. This manipulation of registers is enough to implement a data dependency from the read of x to the write of y.

Abstracting Dependencies. For this lecture however, we're going to model dependencies as fences. This means that the dependencies that we are manipulating here are stronger than in the wild. However the definitions and axioms we're defining should hold with a proper notion of dependencies.

Let's do it! Let's open our `bell` file, and create a tag `dep` for dependencies; we can for example declare a **Fences** type:

```
enum Fences ='dep
```

Now in our litmus test we can use dependencies (note the `f[dep]` instructions in between the read and the write on each thread):

```
Bell LB+dep+dep
{
x = 0;
y = 0;
}
 P0          | P1          ;
 r[] r1 y    | r[] r2 x    ;
 f[dep]      | f[dep]      ;
 w[] x 1     | w[] y 1     ;

exists (0:r1 = 1 /\ 1:r2 = 1)
```

Run it on your current cat file and observe that there are non-SC executions.

Finally we have to give a semantics to our dependencies. The right place to do this is your cat file. Let's look at the execution of the LB litmus test that we give above. We can see a cycle in the union of the program order po and the read-froms rf between threads. We want to build a check such that having dependency arrows instead of program order arrows forbids this execution.

So we need to implement a few concepts in cat speak; let's build the relations yielded by our special dependency events; recall that herd has a standard library in which there is a function called fencerel. This function builds, given a set of events (e.g. F), the relation gathering all pairs of events in program order that have such an event (e.g. f(dep)) between them. Thus using fencerel we can build the relation corresponding to dependencies (to put in your bell file):

```
let deps = fencerel(F & Dep) & (R * _)
show deps
```

Note how we restrict our dependency relation deps to pairs of events that start with a read event (which belongs to the predefined set R). Also, we add show deps to make sure that herd will display this new relation.

Now let's go back to the execution we want to forbid. By the look of it, we want to build a check such that a cycle in the union of the dependencies deps and the external read-froms rfe is forbidden. Let's call this union *happens-before* and write it hb for example. In cat speak (to put in your cat file):

```
let hb = (deps | rfe)+
```

Note that we use the transitive closure ()+ to make hb a transitive relation. We don't really need to for now, especially since we're going to require hb to be acyclic in a minute, but it's going to be important later.

Now let's require, in our cat file, for the happens-before relation to be acyclic to forbid the execution above; in cat speak:

```
procedure no-thin-air() =
  acyclic hb
end
```

Don't forget to call this procedure in your cat file:

```
call no-thin-air()
```

Now run lb+dep+dep.litmus on your current cat file and observe that there is no non-SC execution anymore.

Exercise: Implementing Address Dependencies. How do you think we can implement address dependencies? By this I mean that I'd like to see a sequence of instructions which, placed between two reads, implement e.g. an address dependency.

4.3 Propagation

The third principle is called PROPAGATION. Intuitively, this principle gives the semantics of *fences*, which are special instructions that ensure that two writes of distinct locations separated by a fence have to propagate to other threads in the order in which they appear in the program.

One representative example of this principle is the following:

```
Bell 2+2w
{
x = 0;
y = 0;
}
 P0          | P1          ;
 w[] x 2     | w[] y 2 ;
 w[] y 1     | w[] x 1 ;
exists (x=2 /\ y=2)
```

The 2+2w test has two threads P0 and P1 which both write to the shared memory locations x and y. P0 writes 2 to x and 1 to y, when P2 writes 2 to y and 1 to x. In the end we're asking if it's possible to have both locations holding the value 2, which could be explained by the write-write pairs being reordered on both threads. This corresponds to the following execution, which is non-SC:

2+2w

Try running herd on 2+2w.litmus under our current cat file and observe that there are (flagged) non-SC executions.

Now, placing a fence between each write-write pair should forbid this cycle. Let's do it! Let's open our bell file, and update our fences. Remember that we had defined an annotation 'dep for them previously:

```
enum Fences ='dep
```

Here we additionally give ourselves *lightweight fences* 'lw (there will be heavyweight ones later on):

```
enum Fences ='dep ||'lw
```

Now, placing a lightweight fence between each write-write pair of the 2+2w test (leading to 2+2w+lwfs) should forbid this cycle.

To do so, we first need to define a relation flw, that contains all the possible pairs of events in program order separated by a lightweight fence; recall that we can use the fencerel definition. A good place to do so is our bell file:

```
let flw = fencerel(F & Lw)
show flw
```

Let's also create a **fences** relation to gather all actual fences (as opposed to dependencies); for now we only have lightweight fences, but we'll have heavyweight fences in a moment:

```
let fences = flw
```

Now in our **cat** file we can define the *propagation order* as the order induced by lightweight fences:

```
let prop = fences
```

In cat speak, we can now forbid the 2+2w cycle by the following procedure (to put in your **cat** file):

```
procedure propagation() =
  acyclic prop | co
end
```

And don't forget to call the procedure (in your **cat** file):

```
call propagation()
```

Try running herd on 2+2w.litmus on this new **cat** file and observe that there is no non-SC execution anymore.

4.4   Observation

The fourth principle is called OBSERVATION. Intuitively, this principle means that two reads of distinct locations have to read writes in the order in which these writes propagate. One representative example of this principle is the message passing example we've seen at the beginning:

```
Bell MP
{
x = 0;
y = 0;
}
 P0        | P1         ;
 w[] x 1  | r[] r1 y  ;
 w[] y 1  | r[] r2 x  ;
exists (1:r1 = 1 /\ 1:r2 = 0)
```

As we've seen before, the test as is can yield the result where P1 sees the new flag (r1=1), but reads the stale data (r2=0). There are several architectural reasons for this to happen, amongst which:

1. The two reads on P1 could be reordered—this could happen for example on ARM, IBM Power, or Nvidia machines [3,5];
2. The two writes on P0 could be reordered—this could happen for example on ARM, IBM Power, or Nvidia machines [3,5];
3. The two writes could swap places on their way to the reading thread P1, the write of y by P0 hitting P1 before the write of x does—this could happen for example on ARM or IBM Power machines [5].

We need several devices to protect against each of these items:

1. To protect against the reordering of reads on P1, one typically uses dependencies such as the ones we've defined to deal with LB;
2. To protect against the writes being reordered on their thread or on their way to the reading thread, we need fences.

Now in our litmus test we can use for example:

1. A data dependency to prevent the reordering of reads on P1 (note the f(dep) events in between the reads in the litmus test below);
2. A lightweight fence to prevent the write scenarios we mentioned earlier (note the f(lw) between the writes in the litmus test below).

This corresponds to the following test:

```
Bell MP+lw+dep
{
x = 0;
y = 0;
}
 P0          | P1           ;
 w[] x 1     | r[] r1 y     ;
 f[lw]       | f[dep]       ;
 w[] y 1     | r[] r2 x     ;
exists (1:r1 = 1 /\ 1:r2 = 0)
```

Finally we have to define our OBSERVATION principle. To do so, let's first go back to the LB litmus test: to forbid the non-SC execution, we chose earlier to use dependencies between the read and write on each thread. Equally we could have chosen to place a fence on each thread, like so:

```
Bell LB+lws
{
x = 0;
y = 0;
}
 P0          | P1           ;
 r[] r1 x    | r[] r2 y     ;
 f[lw]       | f[lw]        ;
 w[] y 1     | w[] x 1      ;
exists (0:r1 = 1 /\ 1:r2 = 1)
```

or to mix and match fences and dependencies:

```
Bell LB+dep+lw
{
x = 0;
y = 0;
}
 P0          | P1         ;
 r[] r1 x | r[] r2 y ;
 f[dep]    | f[lw]     ;
 w[] y 1   | w[] x 1   ;
exists (0:r1 = 1 /\ 1:r2 = 1)
```

This means that, in our `cat` file, we can extend our happens-before relation `hb` to include our fences, as follows; `hb` was:

```
let hb = (deps | rfe)+
```

and now becomes:

```
let hb = (deps | fences | rfe)+
```

The `no-thin-air` and `observation` checks can remain unchanged, and now we forbid tests like LB+lws or LB+dep+lw above.

Now, let's look at the execution of the MP litmus test that we want to forbid (note that it is non-SC):

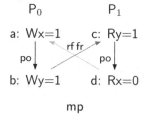

mp

Intuitively, we want to make sure that if the two writes a and b by P0 are separated by a lightweight fence, they cannot be reordered, and propagate to the reading thread P1 in the same order as they appear on P0. Moreover we want to ensure that the two reads c and d by P1 cannot be reordered when separated by a dependency.

Reading off the drawing very plainly, we want to forbid executions where we take one step via a lightweight fence, then one step of read-from, then one step of dependency, then one step of from-read and end up where we started.

Now, observe that there is a `hb` path from the write a by P0 to the read d by P1: $(a, b) \in$ fences, $(b, c) \in$ rf and $(c, d) \in$ deps. It is, however, a special path in `hb`, as its first step consists of a fence step. This special subset of `hb` is in fact the propagation order. Remember that we had defined it in our `cat` file, as follows:

```
let prop = fences
```

which is not quite right anymore, in the light of the MP example. So let's refine our propagation order as follows (to put in your cat file):

```
let prop = fences;hb*
```

Thus the propagation order **prop** starts off with a step of fence, then can continue with a happens-before chain of any length (even zero, as indicated by the reflexive and transitive closure *), through external read-froms, dependencies, and other fences. This is what is sometimes called *B-cumulativity* of a fence [5]: when the propagation order induced by a fence carries over to chains of happens-before.

We can now phrase our OBSERVATION axiom in terms of the propagation order (to put in your cat file):

```
procedure observation() =
  irreflexive fre;prop
end
```

where **fre** is the external from-read, i.e. a from-read arrow between two events that belong to different threads. In cat speak, we can define it using the predefined **ext** relation that gathers pairs of events that belong to different threads, such as the read *d* by P1 and the write *a* by P0. A good place to put the definition of **fre** is in your cat file, for example next to the definition of **rfe**:

```
let fre = fr & ext
```

Don't forget to call the procedure **observation** at the end of your cat file:

```
call observation()
```

Running this new cat file on the tests mp+lw+dep.litmus, lb+lws.litmus and lb+dep+lw.litmus should forbid their non-SC executions.

Observe that the OBSERVATION principle also forbids distributed variants of the message passing example, such as ISA2+lw+dep+dep:

```
Bell ISA2+lw+dep+dep
{
x = 0;
y = 0;
}
 P0          | P1          | P2            ;
 w[] x 1     | r[] r1 y    | r[] r2 z ;
 f[lw]       | f[dep]      | f[dep]     ;
 w[] y 1     | w[] z 1     | r[] r3 x ;
exists (1:r1 = 1 /\ 2:r2 = 1 /\ 2:r3=0)
```

The test ISA2+lw+dep+dep is similar to the message passing one (MP), in that we want to ensure that the two writes by the first thread P0 propagate in the order in which they've been written, to forbid the scenario where, even if the flag y has been passed over to P1, i.e. r1=1, the read of x on P2 reads from

the initial state instead of from the update of x by P0. This corresponds to the
following non-SC execution:

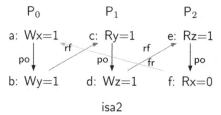

isa2

The difference with MP is that the propagation is over several threads: here
the write of y by P0 propagates to P1, whereas the write of x by P0 propagates
to P2. Because the threads P1 and P2 communicate (via z), and because the
accesses on both P1 and P2 are ordered via dependencies, we have a happens-
before chain from the write of y by P0 to the read of x by P2. This is enough
to create a propagation order arrow from the write of x by P0 to the read of x
by P2, which therefore contradicts the execution where the read of x would read
from the initial state.

The OBSERVATION principle should also forbid WRC+lw+dep:

```
Bell WRC+lw+dep
{
x = 0;
y = 0;
}
 P0        | P1             | P2          ;
 w[] x 1   | r[] r1 x       | r[] r2 y    ;
           | f[lw]          | f[dep]      ;
           | w[] y 1        | r[] r3 x    ;
exists (1:r1=1 /\ 2:r2=1 /\ 2:r3=0)
```

The difference with MP is that the two writes that we want to ensure propagate
in the right order are on different threads: the write of the data x is made by
P0, whereas the write of the flag y is made by P1. We want to ensure that if the
reading thread P2 takes the flag (so that r2=1), then the update of the data x
has reached P2, so that the read of x on P2 cannot read from the initial state
anymore. This would correspond to the following non-SC execution:

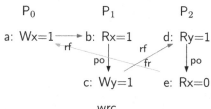

wrc

Our definition of propagation order does not forbid this yet; we have to add
a new notion to our cat file, namely *A-cumulativity*. We say that a fence is

A-cumulative when it orders two writes on different threads P0 and P1 (just like the write a of x by P0 and the write c of y by P1), such that P1 reads the write by P0 then does the fence, then does the second write. Looking at the execution above, it means that an A-cumulative fence placed between b and c should create an arrow between a and c. In cat speak:

```
let A-cumul = rfe;fences
```

Now let's update our propagation order to include A-cumulativity (to put in your cat file):

```
let prop = (fences | A-cumul);hb*
```

Observe that WRC+lw+dep is now forbidden.

Exercise: Distributed. 2+2w Consider the following litmus test, which is essentially a distributed variant of 2+2w:

```
Bell w+rw+ww
{
x = 0;
y = 0;
}
 P0          | P1          | P2          ;
 w[] x 2     | r[] r1 x    | w[] y 2     ;
             | w[] y 1     | w[] x 1     ;
exists (1:r1=2 /\ x=2 /\ y=2)
```

Which fences should we use to forbid the final state? Why? Where should we put them?

4.5   Restoring SC with Heavyweight Fences

The fifth principle explains how to regain SC. To do so, we need to use *heavyweight fences*. We can extend our fences as follows (in our bell file):

```
enum Fences ='dep ||'lw ||'hw
```

Now consider the store buffering litmus test that we've seen earlier:

```
Bell SB
{
x = 0;
y = 0;
}
 P0          | P1          ;
 w[] x 1     | w[] y 1     ;
 r[] r1 y    | r[] r2 x    ;
exists (0:r1 = 0 /\ 1:r2 = 0)
```

Putting a heavyweight fence between the write-read pairs on each thread should forbid the scenario where both reads take their value from the initial state. More generally, putting a heavyweight fence between any pair of events in program order should restore SC.

Recall that SC can be defined as the acyclicity of the union of program order po and the communication relations com. Thus to restore SC with heavyweight fences, we shouldn't allow any cycle in the union of the relation fhw, induced by the heavyweight fences, and the communications.

Let's define fhw in our bell file:

```
let fhw = fencerel(F & Hw)
show fhw
```

And in our cat file, let's implement our fifth principle:

```
procedure restoring-sc() =
  acyclic fhw | com
end
```

```
call restoring-sc()
```

Observe that SB+hws is now forbidden; note however that SB+lws is still allowed, as one really needs heavyweight fences to restore SC. The lightweight fences only contribute to building the propagation order.

Now, heavyweight fences should also forbid the non-SC executions of LB and MP. This means that we should add them to our fences relation, so that they naturally get included in the definitions of hb and prop, thus contribute to the NO-THIN-AIR and OBSERVATION checks. In our cat file, we had:

```
let fences = flw
```

and now we should have:

```
let fences = flw | fhw
```

Think of saving these bell and cat files, for example under the names tiger.bell and tiger.cat.

Exercise: Independent Reads of Independent Writes. Consider the following litmus test, known as IRIW:

```
Bell IRIW
{
x = 0;
y = 0;
}
PO           | P1        | P2        | P3          ;
w[] x 1   | r[] r1 x | w[] y 1 | r[] r3 y ;
          | r[] r2 y |           | r[] r4 x ;
exists (1:r1 = 1 /\ 1:r2 = 0 /\ 3:r3=1 /\ 3:r4=0)
```

Which fences should we use to forbid the final state? Why? Where should we put them?

Exercise: SC and TSO Re-Make Re-Model. [9] Remember that we've defined SC and (not quite) TSO earlier in our `cat` file, as follows:

```
procedure sc() =
  let sc-order = (po | com)+
  acyclic sc-order
end

procedure almost-tso() =
  let ppo = po \ W*R
  let tso-order = ppo | rfe | co | fr
  acyclic tso-order
end
```

Try to reformulate both models in terms of the five principles we've just learnt: SC PER LOCATION, NO THIN AIR, PROPAGATION, OBSERVATION and RESTORING SC. The SC model you'll end up with will be equivalent to the one above. The TSO model you'll end up with will be TSO proper!

5   Let's Herd Our Second Big Cat: A Jaguar [13]

Today we'll learn how to build a model that is similar in spirit to Nvidia GPUs. This model differs from the previous one because GPUs have *scopes*. Start with fresh `bell` and `cat` files (keep the title and include).

Intuitively, a scope is a set of threads. Here we'll consider three different scopes: `cta`, `gpu` and `system`.

5.1   Scopes

Scopes are organised hierarchically: `cta` is *narrower* than `gpu`, and `gpu` is narrower than `system`, as shown in Fig. 2.

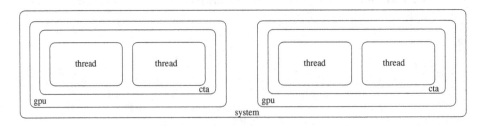

Fig. 2. Concurrency hierarchy

Building the concurrency hierarchy. To build this concurrency hierarchy, the ideal place is our `bell` file; we can simply enumerate these three scopes as follows:

```
enum Scopes ='cta ||'gpu ||'system
```

We also need to specify the hierarchy, with a function called `narrower` (to put in your `bell` file):

```
let narrower(s) = match s with
  ||'system ->'gpu
  ||'gpu ->'cta
end
```

The dual function is called `wider` (to put in your `bell` file):

```
let wider(s) = match s with
  ||'gpu ->'system
  ||'cta ->'gpu
end
```

Scope Tree. Litmus tests also need changing; in particular we need to say which scope a given thread belongs to. Let's go back to the message passing example:

```
Bell MP
{
x = 0;
y = 0;
}
 P0        | P1        ;
 w[] x 1   | r[] r1 y  ;
 w[] y 1   | r[] r2 x  ;
exists (1:r1 = 1 /\ 1:r2 = 0)
```

Let's implement a variant where P0 and P1 are each on a different `cta`, but in the same `gpu`, hence in the same `system`:

```
Bell MP-mit-scopes
{
x = 0;
y = 0;
}
 P0        | P1        ;
 w[] x 1   | r[] r1 y  ;
 w[] y 1   | r[] r2 x  ;
scopes: (system (gpu (cta P0) (cta P1)))

exists (1:r1 = 1 /\ 1:r2 = 0)
```

Note the addition of the *scope tree* `scopes: (system (gpu (cta P0) (cta P1)))` that specifies where the threads P0 and P1 are.

Scope Annotations. Now, we need to say that our instructions can bear these scope annotations, indicating at which level of the concurrency hierarchy they can operate.

In this model, we'd like our fences to have different effects depending on which scope they apply to. To implement that, we can add scope annotations to our fence events (that belong to the predefined set `F`). The right place to do so is in our `bell` file:

```
events F[Scopes]
```

We will need to be able to say, given an instruction or an event of a litmus test, which scope it belongs to. There are two different notions of scoping: *syntactic* and *execution* scopes (sometimes called *static* and *dynamic* [3,7]).

Syntactic Scope. The herd tool provides the primitive `tag2events` which, given a tag such as the scope ones `'cta`, `'gpu` or `'system`, returns all the events that bear this tag. Thus `tag2events('gpu)` returns the set of all events annotated with `'gpu`.

Execution Scope. The herd tool also provides the primitive `tag2scope` which, given a tag, returns the relation that links events that are executed within the same scope level as `tag`. Consider our `MP` example, and let's build `tag2scope ('cta)`: for this particular instance of MP, it will contain eight pairs. The first pair gathers the two writes on P0, because the scope tree `(system (gpu (cta P0) (cta P1)))` specifies that P0 is in its own `cta`. Similarly, the second pair will gather the two reads on P1. The set `tag2scope('cta)` also contains the symmetric pairs to the ones we've just studied, that leads to 4 pairs, to which it adds the four identity pairs, where each event is paired with itself. Note that not all pairs are shown in diagrams, as most relations undergo a transitivity removal procedure before being printed.

Building `tag2scope('gpu)` will gather all the possible pairs of events, because the scope tree specifies that P0 and P1 belong to the same gpu; idem for `tag2scope ('system)`.

5.2   RMO per Scope

So now, experimentally Nvidia GPUs implement *RMO per scope* [3]. Let's study RMO first, then the scopes.

Relaxed Memory Order, or RMO, is a Sparc model that allows the reordering of any pair of read or write events in program order [1,2,11]. One can restore these orderings using dependencies or fences. We've defined the dependencies `deps` earlier. We have defined fences as well, but for today we'll define our fences locally.

Let's write an rmo procedure in our cat file (note that you need the definitions of rfe and fr):

```
procedure rmo() =
  let rmo-fences = fencerel(F)
  let rmo-order = deps | rmo-fences | rfe | co | fr
  acyclic rmo-order
end
```

What we do here is the following: we define a relation fence using our fencerel primitive applied to the set of fence events F. Then we define a relation that we call rmo-order, which is the union of the dependencies deps, the fence relation rmo-fences, the external read-from rfe, the coherence order co, and the from-read fr.

One can show (see e.g. [1,2]) that requiring the acyclicity of this relation rmo-order is enough to implement RMO.

Exercise: Difference Between RMO and Power or ARM. What's a test that distinguishes RMO from Power or ARM (as we've defined them previously)? More precisely, what's a test that's forbidden on RMO but allowed on Power or ARM?

Scope Hierarchy. Now, we need to express the fact that each level of our scope hierarchy (cta, gpu, system) will behave like RMO. To do so, we can modify our rmo procedure to take scopes into account (to put in your cat file):

```
procedure rmo() =
  let rmo-fences(t) = fencerel(F & tag2events(t))
  let rmo-order(t) = (fence(t) | rfe | co | fr) & tag2scope(t)
  forall t in Scopes do
    acyclic rmo-order(t)
  end
end
```

By contrast to our scope-less rmo procedure, here our local relation fence takes a scope annotation t as an argument, and builds the relation induced by fences (which belong to F) that bear the syntactic annotation t. Then the relation rmo-order uses this scoped fence relation; additionally, we impose that the extremities of the pairs gathered in rmo-order belong to the same scope instance of level t. Finally we require the acyclicity of the relation rmo-order for each level of the concurrency hierarchy, i.e. for each t in the set Scopes.

Don't forget to call the procedure rmo from your cat file:

```
call rmo()
```

Now try to run mp-mit-scopes.litmus and observe that it is allowed; try again with the following test:

```
Bell MP-mit-scopes+fgpus
{
x = 0;
y = 0;
}
 P0          | P1            ;
 w[] x 1     | r[] r1 y      ;
 f[gpu]      | f[gpu]        ;
 w[] y 1     | r[] r2 x      ;
scopes: (system (gpu (cta P0) (cta P1)))
```

exists (1:r1 = 1 /\ 1:r2 = 0)

and observe that it is forbidden.

Now think of saving your bell and cat files, for example under the names jaguar.bell and jaguar.cat.

Exercise: Implementing Scope Inclusion. How would you implement an RMO per scope model as above, but where the fences have an effect not only at their scopes, but also at narrower scopes? That is, the fence for system also has an effect at gpu and cta level for example? This procedure should forbid mp-mit-scopes+fgpu+fsys:

```
Bell MP-mit-scopes+fgpu+fsys
{
x = 0;
y = 0;
}
 P0          | P1            ;
 w[] x 1     | r[] r1 y      ;
 f[gpu]      | f[sys]        ;
 w[] y 1     | r[] r2 x      ;
scopes: (system (gpu (cta P0) (cta P1)))
```

exists (1:r1 = 1 /\ 1:r2 = 0)

but not mp-mit-scopes+fcta+fgpu:

```
Bell MP-mit-scopes+fcta+fgpu
{
x = 0;
y = 0;
}
 P0          | P1            ;
 w[] x 1     | r[] r1 y      ;
 f[cta]      | f[gpu]        ;
 w[] y 1     | r[] r2 x      ;
```

```
scopes: (system (gpu (cta P0) (cta P1)))

exists (1:r1 = 1 /\ 1:r2 = 0)
```

6 Let's Herd Our Third Big Cat: A Panther [15]

Today we'll learn how to build a model that is inspired by C++. Start with fresh `bell` and `cat` files (keep the title and include).

This model is different from the previous ones, in particular because it doesn't simply reject executions based on the presence of certain cycles. It also looks for *data races*, and declares an execution that has a data race to be *undefined.*

6.1 Plain and Special Events

Let's first build our `bell` file. We're going to have two different flavours of events: *plain* ones and *special* ones:

```
enum Flavours ='plain ||'special

events R[Flavours]
events W[Flavours]
```

6.2 Release-Acquire Semantics

Now let's focus on our special events. The model we're building is such that synchronisation happens through special events, more precisely, when two threads communicate (i.e. one writes to a location that is read by the other).

Thus the read-from relation over special events is quite central to this model; let's add this notion to our `cat` file, as a relation `special-rf` (for `rf` over special events):

```
let special-rf = rfe & (Special * Special)
```

Now we need to implement the notion that when two threads communicate in an atomic way, they synchronise. Let's build a *happens before* relation `hb`:

```
let hb = (po | special-rf)+
```

Here we say that an event e_1 happens before another one e_2 (i.e. $(e_1, e_2) \in$ hb) when e_1 is in program order before e_2, or e_2 reads from e_1 and they're both special, or any chain of such steps (note how we use the transitive closure).

Note that we didn't say anything specific about fences in our `bell` file; that's because fences won't play much of a role in this model so that we can focus on the special accesses instead. Therefore today in our happens before relation, we're taking all of the program order where we had previously taken only dependencies and fences. On the other hand we're only using the special read-from, whereas previously we used all of `rf`.

Now to make our happens-before relation an order that we can build on, we should implement a NO THIN AIR check. We can put the following procedure in our `cat` file:

```
procedure no-thin-air() =
  acyclic hb
end
```

Now using this happens-before relation we can forbid message passing scenarios from going wrong; recall MP:

```
Bell MP
{
x = 0;
y = 0;
}
 P0          | P1           ;
 w[] x 1  | r[] r1 y  ;
 w[] y 1  | r[] r2 x  ;
exists (1:r1 = 1 /\ 1:r2 = 0)
```

For this test to make sense in our current setup, where reads and writes can be plain or special, we need to annotate our events as being `plain`

```
Bell MP-plain
{
x = 0;
y = 0;
}
 P0                | P1              ;
 w[plain] x 1  | r[plain] r1 y  ;
 w[plain] y 1  | r[plain] r2 x  ;
exists (1:r1 = 1 /\ 1:r2 = 0)
```

If we make the communication over the flag y special, like so:

```
Bell MP-special
{
x = 0;
y = 0;
}
 P0                  | P1                ;
 w[plain] x 1     | r[special] r1 y ;
 w[special] y 1  | r[plain] r2 x   ;
exists (1:r1 = 1 /\ 1:r2 = 0)
```

then we create a happens-before order from the update of the data x (i.e. the write of x on P0) and the read of x on P1.

Now, to forbid the final state of this variant of MP, we need to build an OBSERVATION check (to put in your `cat` file):

```
procedure observation() =
  irreflexive fre;hb
end
```

6.3 Validity

Now let's gather our checks into a single procedure (to put in your `cat` file). Note that we added a call to `sc-per-location cat` files, just because:

```
procedure valid() =
  call sc-per-location()
  call no-thin-air()
  call observation()
end
```

Of course for this to work you need to add the definition of the `sc-per-location` procedure, which you can copy from your previous `cat` files.

Don't forget to call your `valid` procedure (you need `fre` as before):

```
call valid()
```

Now try to run `mp-plain.litmus` under your new `cat` file and observe it is allowed; try it out on `mp-special.litmus` and observe it is forbidden.

6.4 Data Races and Undefined Executions

Now we want to be able to distinguish executions that have data races, and flag them as being undefined. Let's define data races (in our `cat` file):

```
let at-least-one(k) = (k * _ | _ * k)
let conflict = at-least-one(W) & loc & ext
let race =
  let r = conflict & ~(hb | hb^-1)
  in r \ ((I * M) | (M * I) | (Special * Special))
show race
```

Thus we define a race as a pair of accesses that:

- `conflict` (which implies that the accesses are distinct, as they must be on different threads when they conflict), and
- are not ordered by `hb` or `hb^-1` (i.e. take a step of `hb` backwards), and
- none of them is an initialisation write (i.e. belong to the set `I`), and
- are not both special.

A conflict is a pair of accesses, such that at least one is a write (i.e. belongs to W), both accesses are relative to the same memory location (i.e. they belong to loc), and they belong to different threads (i.e. they belong to ext).

Now we can flag racy executions as undefined:

```
procedure race-free() =
  flag ~empty race as undefined
end
```

and define our executions to be both valid and race-free:

```
procedure execution() =
  call valid()
  call race-free()
end
```

Don't forget to call this procedure in your cat file:

```
call execution()
```

and to comment out our previous standalone call to the valid procedure because it appears in our execution procedure now.

Now try out our new cat file on MP-special, and observe that the tool finds it is racy. To fix this issue, we can modify our example like so (beq is a branch instruction; below it branches to END if r1 is equal to 0):

```
Bell MP-special+branch
{
x = 0;
y = 0;
}
 P0                  | P1                 ;
 w[plain] x 1        | r[special] r1 y ;
                     | beq r1, 0, END  ;
 w[special] y 1  | r[plain] r2 x   ;
                     | END:               ;
exists (1:r1 = 1 /\ 1:r2 = 0)
```

This is because the branch on P1 here ensures that the read of x on P1 takes its value only after the flag y has been read by P1. The fact that you need a branch also shows po is a dynamic notion, otherwise hb = (po | rf-special)+ would catch the test without the branch, and not find it racy.

Exercise: Release Sequence. In C++ there's a notion of *release sequence*, which says that synchronisation does not just happen via special-rf, i.e. from the special write of a special read-from to the corresponding special read. Rather, it can happen from any write of the same location and on the same thread that precedes (in program order) a special write.

How would you modify our model to include this notion?

7   Credits

The `cat` language is mostly Luc Maranget's and my work, the five principles as
well [5]. The shiny web interface is thanks to Tyler Sorensen. The `bell` subset has
benefited from Tyler's contribution. Tyler has computed the three pink ponies
at the beginning of this document.

For more ponies: https://youtu.be/3-Y8xLsqywY.

For more `bell` and `cat` files: virginia.cs.ucl.ac.uk/herd-web. In particular:

- IBM Power: http://virginia.cs.ucl.ac.uk/herd-web/?book=herding-cats&lang
  uage=ppc&cat=ppc
- ARM: http://virginia.cs.ucl.ac.uk/herd-web/?book=herding-cats&language=
  arm&cat=arm
- C++: http://virginia.cs.ucl.ac.uk/herd-web/?book=c11popl15
- Nvidia PTX: http://virginia.cs.ucl.ac.uk/herd-web/?book=ptx&language=
  ptx&cat=ptx.

With thanks to the supersonic beta-testers: Patrick Cousot, Matthew Hague,
Luc Maranget, Tyler Sorensen, Michael Tautschnig and Jules Villard. Many
thanks also to Tony Tye for his careful reading.

A   Answers to Exercises

A.1   First Step in Herding Cats

Exercise: what do you think?. Do you think such an execution is possible?
Yes! On IBM Power or ARM machines for example [5].

A.2   First Big Cat: Tiger

Exercise: SC per Location with Load-Load Hazard. Certain architectures, such as Sparc RMO [11] allow what is sometimes called *load-load hazard*, i.e. a situation where the coRR test that we've just seen is allowed to yield the result 0:r1=1; 0:r2=0;. How do you think we can build a check that forbids all tests coWW, coRW1, coRW2 and coWR, but allows the test coRR?

In cat speak:

```
let po-loc-llh = po-loc \ R*R
acyclic po-loc-llh | com as sc-per-location-llh
```

Intuitively what we're doing here is removing the read-read pairs (R*R) from po-loc to build the po-loc-llh relation, then enforcing that this new relation is compatible with the communication relations com.

Let's run herd on coWW, coRW1, coRW2, and coWR, and observe that they are still forbidden. Now let's run it on coRR and observe that it is allowed. We've built a check that enforces SC PER LOCATION but allows for load-load hazards!

Exercise: Implementing Address Dependencies. How do you think we can implement address dependencies? By this I mean that I'd like to see a sequence of instructions which, placed between two reads, implement e.g. an address dependency.

Consider the following variant of MP; the reading thread P1 has an address dependency between its two reads:

```
Bell MP+lw+addr
{
x = 0;
y = 0;
}
P0             | P1            ;
w[] x 1        | r[] r1 y      ;
f(lw)          | xor r3 r1 r1  ;
w[] y 1        | add r4 x r3   ;
               | r[] r2 r4     ;
exists (1:r1 = 1 /\ 1:r2 = 0)
```

This dependency, in conjunction with the lightweight fence on P0, should be enough to forbid the non-SC execution of MP.

Exercise: Distributed. 2+2w Consider the following litmus test, which is essentially a distributed variant of 2+2w:

```
Bell w+rw+ww
{
x = 0;
y = 0;
}
```

```
PO          | P1          | P2          ;
w[] x 2     | r[] r1 x    | w[] y 2     ;
            | w[] y 1     | w[] x 1     ;
exists (1:r1=2 /\ x=2 /\ y=2)
```

Which fences should we use to forbid the final state? Why? Where should we put them?

We should put a lightweight fence between the read-write pair on P1, and a lightweight fence between the write-write pair on P2, like so:

```
Bell w+rw+ww+lws
{
x = 0;
y = 0;
}
PO          | P1          | P2          ;
w[] x 2     | r[] r1 x    | w[] y 2     ;
            | f[lw]       | f[lw]       ;
            | w[] y 1     | w[] x 1     ;
exists (1:r1=2 /\ x=2 /\ y=2)
```

This is because the A-cumulativity of the lightweight fence on P1 will impose an ordering between the write of x on P0 and the write of y on P1.

Exercise: Independent Reads of Independent Writes. Consider the following litmus test, known as IRIW:

```
Bell IRIW
{
x = 0;
y = 0;
}
PO          | P1          | P2          | P3          ;
w[] x 1     | r[] r1 x    | w[] y 1     | r[] r3 y ;
            | r[] r2 y    |             | r[] r4 x ;
exists (1:r1 = 1 /\ 1:r2 = 0 /\ 3:r3=1 /\ 3:r4=0)
```

Which fences should we use to forbid the final state? Why? Where should we put them?

We should put a heavyweight fence between the read-read pairs on P1 and P3, like so:

```
Bell IRIW+hws
{
x = 0;
y = 0;
}
```

```
P0          | P1          | P2          | P3          ;
w[] x 1     | r[] r1 x    | w[] y 1     | r[] r3 y    ;
            | f[hw]       |             | f[hw]       ;
            | r[] r2 y    |             | r[] r4 x    ;
exists (1:r1 = 1 /\ 1:r2 = 0 /\ 3:r3=1 /\ 3:r4=0)
```

This is because IRIW essentially is a distributed variant of the store buffering example, therefore reacts to fences in much the same way as SB, just like ISA2 or WRC react to fences (lightweight in their case) in much the same way as MP.

Thus the A-cumulativity of the heavyweight fence will impose an ordering between the write of x on P0 and the read of y on P1 (idem for the write of y on P2 and the read of x on P3).

Exercise: SC and TSO Re-Make Re-Model [9]. Remember that we've defined SC and (not quite) TSO earlier. Try to reformulate both models in terms of the five principles we've just learnt: SC PER LOCATION, NO THIN AIR, PROPAGATION, OBSERVATION and RESTORING SC. The SC model you'll end up with will be equivalent to the one above. The TSO model you'll end up with will be TSO proper!

Take the tiger cat file, and use the following bell files. For SC:

```
"Re-Make"
let deps = po
let fhw = po
let fences = fhw
```

and for TSO:

```
"Re-Model"

enum Fences = 'wr
events F[Fences]
let fwr = fencerel(F & Wr)
let deps = po
let fhw = fwr | po \ (W*R)
let fences = fhw
show fwr, fhw
```

A.3 Second Big Cat: Jaguar

Exercise: Difference Between RMO and Power or ARM What's a test that distinguishes RMO from Power or ARM (as we've defined them previously)? More precisely, what's a test that's forbidden on RMO but allowed on Power or ARM?

IRIW+deps distinguishes RMO from Power or ARM:

```
Bell IRIW+deps
{
x = 0;
y = 0;
}
P0          | P1          | P2         | P3           ;
w[] x 1     | r[] r1 x    | w[] y 1    | r[] r3 y   ;
            | f[dep]      |            | f[dep]      ;
            | r[] r2 y    |            | r[] r4 x   ;
exists (1:r1 = 1 /\ 1:r2 = 0 /\ 3:r3=1 /\ 3:r4=0)
```

because it is forbidden on RMO [1,2], but allowed on Power and ARM [5].

Exercise: Implementing Scope Inclusion. How would you implement an RMO per scope model as above, but where the fences have an effect not only at their scopes, but also at narrower scopes? That is, the fence for system also has an effect at gpu and cta level for example? This procedure should forbid mp-mit-scopes+fgpu+fsys, but not mp-mit-scopes+fcta+fgpu.

We need to invent a recursive notion of wider, that will return, given a scope level s, *all* scope levels wider than s, not just the immediately wider one:

```
let wider2(s) = match s with
  || 'gpu -> 'system
  || 'cta -> 'gpu | 'system
end
```

We can then use this new notion in the way we define our rmo-fences:

```
procedure rmo() =
  let rmo-fences(t) = fencerel(F & wider2(t))
  let rmo-order(t) = (fence(t) | rfe | co | fr) & tag2scope(t)
  forall t in Scopes do
    acyclic rmo-order(t)
  end
end
```

Note how we replace the call to tag2events by a call to wider2 in the definition of rmo-fences.

Exercise: Release Sequence. In C++ there's a notion of *release sequence*, which says that synchronisation does not just happen via special-rf, i.e. from the special write of a special read-from to the corresponding special read. Rather, it can happen from any write of the same location and on the same thread that precedes (in program order) a special write.

How would you modify our model to include this notion?

Like so (where ? means zero or one step):

```
let hb = (po | co?;special-rf)+
```

B   Complete Bell and Cat Files

B.1   Kittens

Here's `kittens.bell`:

```
"I'm at your service ma'am, and here's your kitty back"
enum Fences = 'wr
events F[{'wr}]

let fwr = fencerel(F & Wr)
show fwr
```

Here's `kittens.cat`:

```
"How can I thank you for your bringing kitty back?"

include "tutorial.cat"

let fr = rf^-1;co
show fr

let rfe = rf & ext

let com = rf | co | fr

(* SC *)

procedure sc() =
  let sc-order = (po | com)+
  acyclic sc-order
end

(* call sc() *)

(* Almost TSO *)

procedure almost-tso() =
  let ppo = po \ W*R
  let tso-order = ppo | fwr | rfe | co | fr
  acyclic tso-order
end

call almost-tso()
```

B.2   Tiger

Here's `tiger.bell`:

```
"He went out tiger hunting with his elephant and gun"

enum Fences = 'dep || 'lw || 'hw
events F[Fences]

let deps = fencerel(F & Dep) & (R * _)
show deps

let flw = fencerel(F & Lw)
show flw

let fhw = fencerel(F & Hw)
show fhw

let fences = flw | fhw
```

Here's `tiger.cat`:

```
"In case of accidents he always took his mum"

include "tutorial.cat"

let fr = rf^-1;co
show fr
let fre = fr & ext
let rfe = rf & ext
let com = rf | co | fr

(* Flag non-SC executions *)

procedure sc-flag() =
  let sc-order = (po | com)+
  flag ~acyclic sc-order as non-sc
end

call sc-flag()

(* SC per location *)

let po-loc = po & loc

procedure sc-per-location() =
```

```
    acyclic po-loc | com
end

call sc-per-location()

(* No thin air *)

let hb = (deps | fences | rfe)+

procedure no-thin-air() =
  acyclic hb
end

call no-thin-air()

(* Propagation *)
let A-cumul = rfe;fences
let prop = (fences | A-cumul);hb*

procedure propagation() =
  acyclic prop | co
end

call propagation()

procedure observation() =
  irreflexive fre;prop
end

call observation()

procedure restoring-sc() =
  acyclic fhw | com
end

call restoring-sc()
```

B.3   Jaguar

Here's jaguar.bell:

"I'll call you Jaguar"

```
enum Scopes = 'cta || 'gpu || 'system
```

```
let narrower(s) = match s with
  || 'system -> 'gpu
  || 'gpu -> 'cta
end

let wider(s) = match s with
  || 'gpu -> 'system
  || 'cta -> 'gpu
end

events F[Scopes]
```

Here's jaguar.cat:

```
"If I may be so bold"

include "tutorial.cat"

let fr = rf^-1;co
let rfe = rf & ext

procedure rmo() =
  let fence(t) = fencerel(F & tag2events(t))
  let rmo-order(t) = (fence(t) | rfe | co | fr) & tag2scope(t)
  forall t in Scopes do
    acyclic rmo-order(t)
  end
end

call rmo()
```

B.4   Panther

Here's panther.bell:

```
"Kool thing walkin' like a panther"

enum Flavours = 'plain || 'special

events R[Flavours]
events W[Flavours]
```

Here's panther.cat:

```
"Come on and give me an answer"
```

```
include "tutorial.cat"

(* SC per location *)

let po-loc = po & loc

let fr = rf^-1;co
let fre = fr & ext
show fr

let com = rf | co | fr

procedure sc-per-location() =
  acyclic po-loc | com
end
(* No thin air *)

let special-rf = rfe & (Special * Special)

let hb = (po | special-rf)+

procedure no-thin-air() =
  acyclic hb
end

(* Observation *)

procedure observation() =
  irreflexive fre;hb
end

(* Valid executions *)

procedure valid() =
  call sc-per-location()
  call no-thin-air()
  call observation()
end

(* call valid() *)

(* Races *)

let at-least-one k = (k * _ | _ * k)
let conflict = at-least-one(W) & loc & ext
```

```
let race =
  let r = conflict & ~(hb | hb^-1)
  in r \ (id | (I * M) | (M * I) | (Special * Special))
show race

procedure race-free() =
  flag ~empty race as undefined
end

procedure execution() =
  call valid()
  call race-free()
end

call execution()
```

References

1. Alglave, J.: A shared memory poetics. Ph.D. thesis, Université Paris 7 (2010)
2. Alglave, J.: A formal hierarchy of weak memory models. Formal Methods Syst. Des. **41**(2), 178–210 (2012)
3. Alglave, J., Batty, M., Donaldson, A.F., Gopalakrishnan, G., Ketema, J., Poetzl, D., Sorensen, T., Wickerson, J.: GPU concurrency: weak behaviours and programming assumptions. In: ASPLOS (2015)
4. Alglave, J., Kroening, D., Lugton, J., Nimal, V., Tautschnig, M.: Soundness of data flow analyses for weak memory models. In: Yang, H. (ed.) APLAS 2011. LNCS, vol. 7078, pp. 272–288. Springer, Heidelberg (2011)
5. Alglave, J., Maranget, L., Tautschnig, M.: Herding cats: modelling, simulation, testing, and data-mining for weak memory. TOPLAS **36**(2) (2014)
6. The Beatles: The Continuing Story of Bungalow Bill. In: White Album (1968)
7. Hower, D.R., Hechtman, B.A., Beckmann, B.M., Gaster, B.R., Hill, M.D., Reinhardt, S.K., Wood, D.A.: Heterogeneous-race-free memory models. In: ASPLOS 14 (2014)
8. Lamport, L.: How to make a multiprocessor computer that correctly executes multiprocess programs. IEEE Trans. Comput. **28**(9), 690–691 (1979)
9. Roxy Music: Re-Make Re-Model. In: Roxy Music (1972)
10. Owens, S., Sarkar, S., Sewell, P.: A better x86 memory model: x86-TSO. In: Berghofer, S., Nipkow, T., Urban, C., Wenzel, M. (eds.) TPHOLs 2009. LNCS, vol. 5674, pp. 391–407. Springer, Heidelberg (2009)
11. SPARC International Inc.: The SPARC Architecture Manual Version 9 (1994)
12. Sparks: Here Kitty. In: Hello Young Lovers (2006)
13. T-Rex: Jeepster. In: Electric Warrior (1971)
14. Vafeiadis, V., Balabonski, T., Chakraborty, S., Morisset, R., Nardelli, F.Z.: Common compiler optimisations are invalid in the C11 memory model and what we can do about it. In: POPL (2015)
15. Sonic Youth: Kool Thing. In: Goo (1990)
16. Led Zeppelin: Black Dog. In: Led Zeppelin IV (1971)

A Gentle Introduction to Multiparty Asynchronous Session Types

Mario Coppo[1], Mariangiola Dezani-Ciancaglini[1], Luca Padovani[1], and Nobuko Yoshida[2(✉)]

[1] Dipartimento di Informatica, Università di Torino, Turin, Italy
[2] Department of Computing, Imperial College London, London, UK
n.yoshida@imperial.au.uk

Abstract. This article provides a gentle introduction to multiparty session types, a class of behavioural types specifically targeted at describing protocols in distributed systems based on asynchronous communication. The type system ensures well-typed processes to enjoy non-trivial properties, including communication safety, protocol fidelity, as well as progress. The adoption of multiparty session types can positively affect the whole software lifecycle, from design to deployment, improving software reliability and reducing its development costs.

1 Introduction

In modelling distributed systems where processes interact by means of message passing, one soon realises that many interactions are meant to occur within the scope of private channels following disciplined protocols. We call such private interactions *sessions* and the protocols that describe them *session types*. In its simplest form, a session is established between *two* peers, such as a client connecting with a server. In these cases, the sessions are "binary" or "dyadic" [39, 42]. In general, a session may involve any (usually fixed, but sometimes variable) number of peers. In these cases, we speak of *multiparty sessions* and of their protocol descriptions as of *multiparty session types* [43].

The ability to describe complex interaction protocols by means of a formal, simple and yet expressive type language can have a profound impact on the way distributed systems are designed and developed. This is witnessed by the fact that some important standardisation bodies for web-based business and finance protocols [68, 72, 73] have recently investigated design and implementation frameworks for specifying message exchange rules and validating business logic based on the notion of *multiparty sessions*, where multiparty session types are "shared agreements" between teams of programmers developing possibly large and complex distributed protocols or software systems.

A multiparty session type theory consists of three main ingredients:

- At the most abstract level is the *global type*, which describes a communication protocol from a neutral viewpoint in terms of the interactions that are supposed to occur between the protocol peers, of the order of these interactions, and of the kind of messages exchanged during these interactions.

© Springer International Publishing Switzerland 2015
M. Bernardo and E.B. Johnsen (Eds.): SFM 2015, LNCS 9104, pp. 146–178, 2015.
DOI: 10.1007/978-3-319-18941-3_4

- At the most concrete level are *processes*, which describe the behaviour of the peers involved in the session using a formal language (usually, a dialect of the π-calculus).
- Somehow in between these two levels are *local types*, one for each peer, which describe the same communication protocol as the global type, but from the viewpoint of each peer.

These ingredients are strictly related: a *projection operation* extracts the local type of each peer from the global type, and a *type system* makes sure that a process uses the communication channels it owns according to their local type. Once these relations are established, a number of properties can be proved, among which:

- *communication safety*, namely the fact that there is never a mismatch between the types of sent and expected messages, despite the same communication channel is used for exchanging messages of different types;
- *protocol fidelity*, namely the fact that the interactions that occur are accounted for by the global type and therefore are allowed by the protocol;
- *progress*, namely the fact that every message sent is eventually received, and every process waiting for a message eventually receives one.

Remarkably, these properties are guaranteed by means of purely local checks on the single peers that participate in the protocol, despite the fact that they will run independently once the session has been established. The ability to prove relevant global properties by means of local checks is one of the key features of session type theories.

The present article formalises these concepts and provides a gentle introduction to multiparty session type theory. The process calculus and the type system we use have been first introduced in [3] and then developed in [25]. Notably, the focus of these two papers was the design of a type system assuring progress even in presence of session interleaving. In this article we solely describe the so-called *communication type system*, which assures communication safety, protocol fidelity and, when no sessions are interleaved, progress.

Outline. We start illustrating our calculus with simple yet comprehensive examples in Sect. 2. The calculus of asynchronous, multiparty sessions is the content of Sect. 3. The communication type system assuring that processes behave correctly with respect to the sessions in which they are involved is illustrated with examples in Sect. 4. Section 5 discusses related work and further readings. To ease reading and accessibility of the content, proofs of the properties enjoyed by well-typed processes and additional technical material have been collected in the Appendix.

2 Examples

In this section we present two versions of a simple but non-trivial example that illustrates the basic functionalities and features of the process calculus that we work with. This example comes from a Web service usecase in Web Service Choreography Description Language (WS-CDL) Primer 1.0 [73], capturing a collaboration pattern typical to many business and distributed protocols [62,69,72].

2.1 Example 1: The Three Buyer Protocol

The setting is that of a system involving Alice, Bob, and Carol that cooperate in order to buy a book from a Seller. The participants follow a protocol that is described informally below:

1. Alice sends a book title to Seller and Seller sends back a quote to Alice and Bob. Alice tells Bob how much she can contribute.
2. If the price is within Bob's budget, Bob notifies both Seller and Alice he accepts, then sends his address to Seller and Seller answers with the delivery date.
3. If the price exceeds Bob's budget, Bob asks Carol to collaborate by establishing a new session. Bob sends Carol how much she has to contribute and *delegates* the remaining interactions with Alice and Seller to her.
4. If Carol's contribution is within her budget, she accepts the quote, notifies Alice, Bob and Seller, and continues the rest of the protocol with Seller and Alice *as if she were Bob*. Otherwise, she notifies Alice, Bob and Seller to quit the protocol.

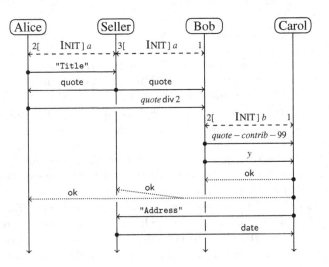

Fig. 1. An execution of the three buyer protocol.

Figure 1 depicts an execution of the above protocol where Bob asks Carol to collaborate (by delegating the remaining interactions with Alice and Seller) and the transaction terminates successfully.

Multiparty session programming consists of two steps: specifying the intended communication protocols using global types and implementing these protocols using processes. The specifications of the three-buyer protocol are given as two distinct global types: one is G_a among Alice, Bob and Seller and the other is G_b between Bob and Carol. In G_a Alice plays role 2, Bob plays role 1, and Seller plays role 3, while in G_b Bob plays role 2 and Carol plays role 1. We annotate the global types with line numbers (i) so that we can easily refer to the actions in them.

$$G_a =$$
$$(1) \quad 2 \longrightarrow 3 : \quad \langle string \rangle.$$
$$(2) \quad 3 \longrightarrow \{1,2\} : \langle int \rangle.$$
$$(3) \quad 2 \longrightarrow 1 : \quad \langle int \rangle.$$
$$(4) \quad 1 \longrightarrow \{2,3\} : \{ok : \quad 1 \longrightarrow 3 : \langle string \rangle.$$
$$(5) \qquad\qquad\qquad\qquad\qquad 3 \longrightarrow 1 : \langle date \rangle.end,$$
$$(6) \qquad\qquad\qquad\qquad quit : \quad end\}$$

$$G_b =$$
$$(1) \quad 2 \longrightarrow 1 : \langle int \rangle.$$
$$(2) \quad 2 \longrightarrow 1 : \langle T \rangle.$$
$$(3) \quad 1 \longrightarrow 2 : \{ok : end, quit : end\}$$

$$T = \oplus \langle \{2,3\}, \{ok : !\langle 3, string \rangle.?(3, date).end, quit : end\} \rangle$$

Global types provide an overall description of the two conversations, directly abstracting the scenario of the diagram. In G_a, line (1) denotes Alice sending a string value to Seller. Line (2) says that Seller sends the same integer value to Alice and Bob and line (3) says that Alice sends an integer to Bob. In lines (4-6) Bob sends either ok or quit to Seller and Alice. In the first case Bob sends a string to Seller and receives a date from Seller, in the second case there are no further communications.

Line (2) in G_b represents the delegation of a channel with the communication behaviour specified by the session type T from Bob to Carol (note that Seller and Alice in T concern the session on a). Then Carol terminates the interaction as if she were Bob in session a. Note that in this case the Seller does not know if he is talking with Bob or Alice.

Table 1 shows an implementation of the three buyer protocol conforming to G_a and G_b for the processes Seller, Alice, Bob, and Carol in the calculus that we will formally define in Sect. 3.1. The service name a is used for initiating sessions corresponding to the global type G_a. Seller initiates a three party session by means of the session request operation $\overline{a}[3](y)$, where the index 3 identifies Seller. Since 3 is also the overall number of participants in this session, a occurs with an over-bar. Alice and Bob get involved in the session by means of the session accept operations $a[1](y)$ and $a[2](y)$ and the indexes 2 and 1 identify them as Alice and Bob, respectively. Once the session has started, Seller, Alice and Bob communicate using their private channels represented by y. Each channel y can be interpreted as a session endpoint connecting a participant with all the others in the same session; the receivers of the data sent on y are specified by giving the participant numbers. Line (1) of G_a is implemented by the matching output and input actions $y!\langle p, \texttt{"Title"} \rangle$ of Alice and $y?(2, title)$ of the Seller.

Table 1. Implementation of the three buyer protocol.

Seller $= \overline{a}[3](y).y?(2, title).y!\langle\{1,2\}, quote\rangle.y\&(1, \{\text{ok} : y?(1, address).y!\langle 1, date\rangle.\mathbf{0}, \text{quit} : \mathbf{0}\})$

Alice $= a[2](y).y!\langle 3, \texttt{"Title"}\rangle.y?(3, quote)).y!\langle 1, quote \text{ div } 2\rangle.y\&(1, \{\text{ok} : \mathbf{0}, \text{ quit} : \mathbf{0}\})$

Bob $= a[1](y).y?(3, quote).y?(2, contrib).\text{if } (quote - contrib < 100)$
$$\text{then } y\oplus\langle\{2,3\}, \text{ok}\rangle.y!\langle 3, \texttt{"Address"}\rangle.y?(3, date).\mathbf{0}$$
$$\text{else}\overline{b}[2](z).z!\langle 1, quote - contrib - 99\rangle.z!\langle\langle 1, y\rangle\rangle.z\&(1, \{\text{ok} : \mathbf{0}, \text{quit} : \mathbf{0}\})$$

Carol $= b[1](z).z?(2, x).z?((2, t)).\text{if } (x < 100)$
$$\text{then } z\oplus\langle 2, \text{ok}\rangle.t\oplus\langle\{2,3\}, \text{ok}\rangle.t!\langle 3, \texttt{"Address"}\rangle.t?(3, date).\mathbf{0}$$
$$\text{else } z\oplus\langle 2, \text{quit}\rangle.t\oplus\langle\{2,3\}, \text{quit}\rangle.\mathbf{0}$$

Line (2) of G_a is implemented by the output action $y!\langle\{1,2\}, \text{quote}\rangle$ of the Seller which is matched by the input actions $y?(3, quote)$ of both Bob and Alice. Line (3) of G_b is implemented by the selection and branching actions $z\oplus\langle 2, \text{ok}\rangle$, $z\oplus\langle 2, \text{quit}\rangle$ and $z\&(1, \{\text{ok} : \mathbf{0}, \text{quit} : \mathbf{0}\})$.

In process Bob, if the quote minus Alice's contribution exceeds 100, another session between Bob and Carol is established through the shared service name b. Delegation occurs by passing the private channel y from Bob to Carol (actions $z!\langle\langle 1, y\rangle\rangle$ and $z?((2, t))$), so that the rest of the session with Seller and Alice is carried out by Carol.

In this particular example no deadlock is possible, even if different sessions are interleaved with each other and the communication topology changes because of delegation.

2.2 Example 2: The Three Buyer Protocol with Recursion

We now describe a variant of the above example that uses recursion. The scenario is basically the same, the only part that changes is that, if the price exceeds the budget, Bob initiates a negotiation with Carol to collaborate together by establishing a new session: Bob starts asking a first proposal of contribution to Carol. At each step Carol answers with a new offer. Bob can accept the offer, try with a new proposal or give up. When Bob decides to end the negotiation (accepting the offer or giving up) he communicates the exit to Carol and, as before, Carol concludes the protocol with Seller.

Figure 2 depicts the part of the protocol involving recursion.

The communication protocols are described by the following global types; these are similar to the ones of the previous example. In particular G_a is exactly the same (since the server does not notice the further interactions among the buyers). Instead, G_b is now more involved since we have a recursive part which represents the (possibly) recursive negotiation between Bob and Carol.

$G_a =$

$\begin{pmatrix}1\\2\\3\\4\\5\\6\end{pmatrix}$
1. $2 \longrightarrow 3 :$ $\langle string \rangle.$
2. $3 \longrightarrow \{1,2\} : \langle int \rangle.$
3. $2 \longrightarrow 1 :$ $\langle int \rangle.$
4. $1 \longrightarrow \{2,3\} : \{ok : 1 \longrightarrow 3 : \langle string \rangle.$
5. $3 \longrightarrow 1 : \langle date \rangle.end,$
6. $quit : end\}$

$G_b =$

$\begin{pmatrix}1\\2\\3\end{pmatrix}$
1. $\mu t.2 \longrightarrow 1 : \langle int \rangle.$
2. $1 \longrightarrow 2 : \langle int \rangle.$
3. $2 \longrightarrow 1 : \{ok : 2 \longrightarrow 1 : \langle T \rangle.end,$
     $more : t,$
     $quit : end\}$

$$T = \oplus \langle \{3,2\}, \{ok : !\langle 3, string \rangle.?(3, date).end, quit : end\}\rangle$$

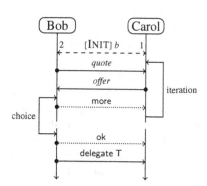

Fig. 2. The three buyer protocol with recursion: additional interactions between Bob and Carol.

$\text{Seller} = \overline{a}[3](y).y_3?(2, title).y!\langle\{1,2\}, quote\rangle.y\&(1, \{ok : y?(1, address).y!\langle 1, date\rangle.0, \ quit : 0\})$

$\text{Alice} = a[2](y).y!\langle 3, \texttt{"Title"}\rangle.y?(3, quote).y!\langle 1, quote \ div \ 2\rangle.y\&(1, \{ok : 0, \ quit : 0\})$

$\text{Bob} = a[1](y).y?(3, quote).y?(2, contrib).$
    if $(quote \text{ - } contrib < 100)$ then $y \oplus \langle\{2,3\}, ok\rangle.y!\langle 3, \texttt{"Address"}\rangle.y?(3, date).0$
    else $\overline{b}[2](z).$
        def $X(x', z', y') =$
            $z'!\langle 1, x'\rangle.z?(1, w).$
            if $good(w)$ then $z' \oplus \langle 1, ok\rangle.z'!\langle\langle 1, y'\rangle\rangle.0$
            else if $negotiable(w)$ then $z' \oplus \langle 1, more\rangle.X\langle newproposal(w), z', y'\rangle$
                else $z' \oplus \langle 1, quit\rangle.y' \oplus \langle\{2,3\}, quit\rangle.0$
    in $X\langle firstproposal(quote), z, y\rangle$

$\text{Carol} = b[1](z).$ def $Y(z') = z'?(2, x).z'!\langle 2, offer(x)\rangle.$
        $z'\&(2, \{ok : z'?((2, t)).t \oplus \langle\{2,3\}, ok\rangle.t!\langle 3, \texttt{"Address"}\rangle.t?(3, date).0$
            $more : Y\langle z'\rangle,$
            $quit : 0\})$
    in $Y\langle z'\rangle$

Fig. 3. The three buyer example with recursion.

The code of the example is in Fig. 3. Again, it is similar to the previous one, but for the recursive definitions in the processes Bob and Carol. Note that the recursive process X in Bob's code has a data parameter (x') and two channel parameters (y' and z'), while the process Y in Carol's code has only one channel parameter (z').

3   The Calculus for Multiparty Sessions

In this section we formalise syntax and operational semantics of the calculus of multiparty asynchronous sessions. To ease the presentation and limit some technicalities, with respect to the previous section we consider a slightly simpler calculus in which communication actions always specify exactly one receiver instead of a non-empty set of receivers and we assume that recursive definitions have exactly one data parameter and one channel parameter. Allowing multiple receivers is mostly a matter of syntactic sugar, since a communication action involving multiple receivers can be canonically encoded as a sequence of actions involving single receivers only. Some notions, however, such as the projection operator, are affected by this design choice and should be adjusted accordingly. The interested reader may refer to [25] for the presentation of the calculus with native support for multiple receivers, and to [34] for the definition of a projection operator that can handle actions with multiple receivers encoded as sequences of actions with single receivers only.

3.1   Syntax

The present calculus is a variant of the calculus in [43], as explained in Sect. 5. The syntax of *processes*, ranged over by $P, Q \ldots$, and that of *expressions*, ranged over by $e, e', \ldots$, is given by the grammar in Table 2, which shows also naming conventions. The operational semantics is defined by a set of reduction rules. In the reduction of processes it is handy to introduce elements, like queues of messages and runtime channels, which are not expected to occur in the source code written by users (*user processes*). These elements, which are referred as *runtime syntax*, appear shaded .

The processes of the form $\overline{u}[\mathsf{p}](y).P$ and $u[\mathsf{p}](y).P$ cooperate in the initiation of a multiparty session through a service name identified by u, where p denotes a *participant* to the session. Participants are represented by progressive numbers and are ranged over by $\mathsf{p}, \mathsf{q}, \ldots$ The barred identifier is the one corresponding to the participant with the highest number, which also gives the total number of participants needed to start the session. The (bound) variable y is the placeholder for the channel that will be used in the communications. After opening a session each channel placeholder will be replaced by a *channel with role* $s[\mathsf{p}]$, which represents the runtime channel of the participant p in the session s.

Process communications (communications that can only take place inside initiated sessions) are performed using the next three pairs of primitives: the sending and receiving of a value; the channel delegation and reception (where the process performing the former action delegates to the process receiving it the capability to participate in a session by passing a channel associated with that session); and the selection and branching (where the former action sends one

Table 2. Process syntax and naming conventions.

$$
\begin{aligned}
P ::= &\ \overline{u}[\mathbf{p}](y).P && \text{Multicast request} \\
\mid &\ u[\mathbf{p}](y).P && \text{Accept} \\
\mid &\ c!\langle \mathbf{p}, e\rangle.P && \text{Value sending} \\
\mid &\ c?(\mathbf{p}, x).P && \text{Value reception} \\
\mid &\ c!\langle\!\langle \mathbf{p}, c'\rangle\!\rangle.P && \text{Channel delegation} \\
\mid &\ c?((\mathbf{q}, y)).P && \text{Channel reception} \\
\mid &\ c \oplus \langle \mathbf{p}, l\rangle.P && \text{Selection} \\
\mid &\ c\&(\mathbf{p}, \{l_i : P_i\}_{i \in I}) && \text{Branching} \\
\mid &\ \text{if } e \text{ then } P \text{ else } Q && \text{Conditional} \\
\mid &\ P \mid Q && \text{Parallel} \\
\mid &\ \mathbf{0} && \text{Inaction} \\
\mid &\ (\nu a)P && \text{Service name hiding} \\
\mid &\ \text{def } D \text{ in } P && \text{Recursion} \\
\mid &\ X\langle e, c\rangle && \text{Process call} \\
\mid &\ (\nu s)P && \text{Session hiding} \\
\mid &\ s : h && \text{Message queue} \\
D ::= &\ X(x, y) = P && \text{Declaration} \\
\mathscr{E} ::= &\ [\,] \mid P \mid (\nu a)\mathscr{E} && \text{Evaluation context} \\
\mid &\ (\nu s)\mathscr{E} \mid \text{def } D \text{ in } \mathscr{E} \\
\mid &\ \mathscr{E} \mid \mathscr{E}
\end{aligned}
$$

$$
\begin{aligned}
a, b && \text{Service name} \\
x && \text{Value variable} \\
y, z, t && \text{Channel Variable} \\
s && \text{Session name} \\
\mathbf{p}, \mathbf{q} && \text{Participant number} \\
X, Y && \text{Process variable} \\
l && \text{Label} \\
s[\mathbf{p}] && \text{Channel with role} \\
u ::= x \mid a && \text{Identifier} \\
v ::= a \mid \text{true} && \text{Value} \\
\mid \text{false} \\
e ::= v \mid x \\
\mid e \text{ and } e' && \text{Expression} \\
\mid \text{not } e \dots \\
c ::= y \mid s[\mathbf{p}] && \text{Channel} \\
m ::= (\mathbf{q}, \mathbf{p}, v) && \text{Message in transit} \\
\mid (\mathbf{q}, \mathbf{p}, s[\mathbf{p}']) \\
\mid (\mathbf{q}, \mathbf{p}, l) \\
h ::= h \cdot m \mid \varnothing && \text{Queue}
\end{aligned}
$$

of the labels offered by the latter). The input/output operations (including the delegation ones) specify the channel and the sender or the receivers, respectively. Thus, $c!\langle \mathbf{p}, e\rangle$ denotes the sending of a value on channel c to the participant $\mathbf{p}$; accordingly, $c?(\mathbf{p}, x)$ denotes the intention of receiving a value on channel c from the participant $\mathbf{p}$. The same holds for delegation/reception (but the receiver is only one) and selection/branching.

An *output action* is a value sending, channel delegation or label selection: an *output process* is a process whose first action is an output action. An *input action* is a value reception, session reception or label branching: an *input process* is a process whose first action is an input action. A *communication action* is either an output or an input action.

As usual evaluation contexts are processes with some holes.

As in [43], we use message queues in order to model TCP-like asynchronous communications (where message order is preserved and sending is non-blocking). A message in a queue can be a value message, $(\mathbf{q}, \mathbf{p}, v)$, indicating that the value v was sent by the participant $\mathbf{q}$ and the recipients is the participant $\mathbf{p}$; a channel message (delegation), $(\mathbf{q}, \mathbf{p}, s[\mathbf{p}'])$, indicating that $\mathbf{q}$ delegates to $\mathbf{p}$ the role of $\mathbf{p}'$ on the session s (represented by the channel with role $s[\mathbf{p}']$); and a label message, $(\mathbf{q}, \mathbf{p}, l)$ (similar to a value message). The empty queue is denoted by ϕ. By $h \cdot m$ we denote the queue obtained by concatenating m to the queue h. With some abuse of notation we will also write $m \cdot h$ to denote the queue with head element m. By $s : h$ we denote the queue h of the session s. Queues and channels with role are generated by the operational semantics (described later).

We call *pure* a process which does not contain message queues.

There are many binders: request/accept actions bind channel variables, value receptions bind value variables, channel receptions bind channel variables, declarations bind value and channel variables, recursions bind process variables, hidings bind service and session names. In $(\nu s)P$ all occurrences of $s[\mathrm{p}]$ and the queue s inside P are bound. We say that a process is *closed* if the only free names in it are service names (i.e. if it does not contain free variables or free session names).

3.2 Operational Semantics

Processes are considered modulo structural equivalence, denoted by $\equiv$, and defined adding α-conversion to the rules in Table 3. We denote by $\mathrm{fn}(Q)$ $(\mathrm{fn}(D))$ the set of free names in Q (D), by $\mathrm{dpv}(D)$ the set of process variables declared in D and by $\mathrm{fpv}(Q)$ the set of process variables which occur free in Q. Besides the standard rules [50], we have a rule for rearranging messages in a queue when the senders or the receivers are not the same.

Table 3. Structural equivalence.

$$P \mid \mathbf{0} \equiv P \quad P \mid Q \equiv Q \mid P \quad (P \mid Q) \mid R \equiv P \mid (Q \mid R)$$

$$(\nu r)P \mid Q \equiv (\nu r)(P \mid Q) \quad \text{if } r \notin \mathrm{fn}(Q)$$

$$(\nu r)(\nu r')P \equiv (\nu r')(\nu r)P \quad (\nu a)\mathbf{0} \equiv \mathbf{0} \quad (\nu s)(s : \varnothing) \equiv \mathbf{0}$$

$$\text{where } r ::= a \mid s$$

$$\text{def } D \text{ in } \mathbf{0} \equiv \mathbf{0} \quad \text{def } D \text{ in } (\nu r)P \equiv (\nu r)\text{def } D \text{ in } P \quad \text{if } r \notin \mathrm{fn}(D)$$

$$(\text{def } D \text{ in } P) \mid Q \equiv \text{def } D \text{ in } (P \mid Q) \quad \text{if } \mathrm{dpv}(D) \cap \mathrm{fpv}(Q) = \emptyset$$

$$\text{def } D \text{ in } (\text{def } D' \text{ in } P) \equiv \text{def } D' \text{ in } (\text{def } D \text{ in } P)$$
$$\text{if } (\mathrm{dpv}(D) \cup \mathrm{fpv}(D)) \cap \mathrm{dpv}(D') = \mathrm{dpv}(D) \cap (\mathrm{dpv}(D') \cup \mathrm{fpv}(D')) = \emptyset$$

$$s : h \cdot (\mathrm{q},\mathrm{p},\zeta) \cdot (\mathrm{q}',\mathrm{p}',\zeta') \cdot h' \equiv s : h \cdot (\mathrm{q}',\mathrm{p}',\zeta') \cdot (\mathrm{q},\mathrm{p},\zeta) \cdot h' \quad \text{if } \mathrm{p} \neq \mathrm{p}' \text{ or } \mathrm{q} \neq \mathrm{q}'$$

Table 4 shows the reduction rules of processes (we use $\longrightarrow^*$ and $\longrightarrow^k$ with the expected meaning). Rule [Init] describes the initiation of a new session among n participants that synchronise over the service name a. The last participant $\overline{a}[n](y).P_n$, distinguished by the overbar on the service name, specifies the number n of participants. After the initiation, the participants will share the private session name s, and the queue associated to s, which is initially empty. The variable y in each participant p will be replaced by the corresponding channel with role $s[\mathrm{p}]$. The output rules [Send], [Deleg] and [Sel] enqueue values, channels and labels, respectively, into the queue of the session s (in rule [Send], $e \downarrow v$ denotes the evaluation of the expression e to the value v). The input rules [Rcv], [SRcv] and [Branch] perform the corresponding complementary operations. Note that

Table 4. Reduction rules.

$$a[1](y).P_1 \mid ... \mid a[n-1](y).P_{n-1} \mid \overline{a}[n](y).P_n \longrightarrow$$
$$(\nu s)(P_1\{s[1]/y\} \mid ... \mid P_{n-1}\{s[n-1]/y\} \mid P_n\{s[n]/y\} \mid s : \varnothing) \qquad \text{[Init]}$$

$$s[\mathsf{p}]!\langle \mathsf{q},e\rangle.P \mid s:h \longrightarrow P \mid s:h \cdot (\mathsf{p},\mathsf{q},v) \quad (e{\downarrow}v) \qquad \text{[Send]}$$

$$s[\mathsf{p}]!\langle\langle \mathsf{q},s'[\mathsf{p}']\rangle\rangle.P \mid s:h \longrightarrow P \mid s:h \cdot (\mathsf{p},\mathsf{q},s'[\mathsf{p}']) \qquad \text{[Deleg]}$$

$$s[\mathsf{p}] \oplus \langle l,\mathsf{q}\rangle.P \mid s:h \longrightarrow P \mid s:h \cdot (\mathsf{p},\mathsf{q},l) \qquad \text{[Sel]}$$

$$s[\mathsf{p}]?(\mathsf{q},x).P \mid s:(\mathsf{q},\mathsf{p},v) \cdot h \longrightarrow P\{v/x\} \mid s:h \qquad \text{[Rcv]}$$

$$s[\mathsf{p}]?((\mathsf{q},y)).P \mid s:(\mathsf{q},\mathsf{p},s'[\mathsf{p}']) \cdot h \longrightarrow P\{s'[\mathsf{p}']/y\} \mid s:h \qquad \text{[SRcv]}$$

$$s[\mathsf{p}]\&(\mathsf{q},\{l_i:P_i\}_{i\in I}) \mid s:(\mathsf{q},\mathsf{p},l_j) \cdot h \longrightarrow P_j \mid s:h \quad (j \in I) \qquad \text{[Branch]}$$

$$\text{if } e \text{ then } P \text{ else } Q \longrightarrow P \quad (e{\downarrow}\text{true}) \quad \text{if } e \text{ then } P \text{ else } Q \longrightarrow Q \quad (e{\downarrow}\text{false}) \qquad \text{[If-T, If-F]}$$

$$\text{def } X(x,y) = P \text{ in } (X\langle e,s[\mathsf{p}]\rangle \mid Q) \longrightarrow \text{def } X(x,y) = P \text{ in } (P\{v/x\}\{s[\mathsf{p}]/y\} \mid Q) \quad (e{\downarrow}v) \ \text{[ProcCall]}$$

$$P \longrightarrow P' \quad \Rightarrow \quad \mathcal{E}[P] \longrightarrow \mathcal{E}[P'] \qquad \text{[Ctxt]}$$

$$P \equiv P' \text{ and } P' \longrightarrow Q' \text{ and } Q \equiv Q' \quad \Rightarrow \quad P \longrightarrow Q \qquad \text{[Str]}$$

these operations check that the sender matches, and also that the message is actually meant for the receiver.

4 Communication Type System

This section introduces the communication type system, by which we can check type soundness of the communications and protocol fidelity. This type system is the one introduced in [25], but the proof of subject reduction is cleaned up by the use of the property stated in Lemma 1. As we have done in Sect. 3.1, here too we only consider communication actions with single receivers, even though the examples make use of a slightly more general syntax.

4.1 Global and Session Types

Global types describe the whole conversation scenarios of multiparty session. *Session types* correspond to projections of global types on the individual participants: they are types of pure processes. The grammar for global and session types is given in Table 5. *Sorts* $S, S', \dots$ are associated to values (either base types or *closed* global types, ranged over by G). *Exchange types* $U, U', \dots$ consist of sort types or *closed* session types, ranged over by T.

The global type $\mathsf{p} \to \mathsf{q} : \langle S\rangle.G$ says that participant p sends a value of sort S to the participant $\mathsf{q} \neq \mathsf{p}$ and then the interactions described in G take place.

Table 5. Global and session types.

$$
\begin{array}{lll}
S & ::= & \textsf{bool} \mid \dots \mid G \qquad \text{Sorts} \\
U & ::= & S \mid \mathsf{T} \qquad\qquad\quad \text{Exchange types}
\end{array}
$$

Global types

$$
\begin{array}{lll}
G & ::= & \mathsf{p} \to \mathsf{q} : \langle S \rangle.G \qquad \text{Value exchange} \\
 & \mid & \mathsf{p} \to \mathsf{q} : \langle \mathsf{T} \rangle.G \qquad \text{Channel exchange} \\
 & \mid & \mathsf{p} \to \mathsf{q} : \{l_i : G_i\}_{i \in I} \;\; \text{Branching} \\
 & \mid & \mu \mathbf{t}.G \mid \mathbf{t} \mid \textsf{end} \qquad \text{Recursion/end}
\end{array}
$$

Session types

$$
\begin{array}{lll}
T & ::= & !\langle \mathsf{p}, S \rangle.T \qquad\qquad \text{Send value} \\
 & \mid & !\langle \mathsf{p}, \mathsf{T} \rangle.T \qquad\qquad \text{Send channel} \\
 & \mid & ?(\mathsf{p}, U).T \qquad\qquad \text{Receive} \\
 & \mid & \oplus \langle \mathsf{p}, \{l_i : T_i\}_{i \in I} \rangle \;\; \text{Selection} \\
 & \mid & \& (\mathsf{p}, \{l_i : T_i\}_{i \in I}) \;\; \text{Branching} \\
 & \mid & \mu \mathbf{t}.T \mid \mathbf{t} \mid \textsf{end} \qquad \text{Recursion/end}
\end{array}
$$

Similarly, the global type $\mathsf{p} \to \mathsf{q} : \langle \mathsf{T} \rangle.G$ says that participant $\mathsf{p} \neq \mathsf{q}$ delegates a channel of type T to participant q and the interaction continues according to G. When it does not matter we use $\mathsf{p} \to \mathsf{q} : \langle U \rangle.G$ to refer both to $\mathsf{p} \to \mathsf{q} : \langle S \rangle.G$ and $\mathsf{p} \to \mathsf{q} : \langle \mathsf{T} \rangle.G$.

Type $\mathsf{p} \to \mathsf{q} : \{l_i : G_i\}_{i \in I}$ says that participant p sends one of the labels l_i to participants q. If l_j is sent, interactions described in G_j take place. Type $\mu \mathbf{t}.G$ is a recursive type, assuming type variables $(\mathbf{t}, \mathbf{t}', \dots)$ are guarded in the standard way, i.e., type variables only appear under some prefix. We take an *equi-recursive* view of recursive types, not distinguishing between $\mu \mathbf{t}.G$ and its unfolding $G\{\mu \mathbf{t}.G/\mathbf{t}\}$ [66, Sect. 21.8]. Type $\textsf{end}$ represents the termination of the session.

Session types represent the input-output actions performed by single participants. The *send types* $!\langle \mathsf{p}, S \rangle.T$, $!\langle \mathsf{p}, \mathsf{T} \rangle.T$ express, respectively, the sending of a value of sort S to participant p or the sending of a channel of type T to participant p followed by the communications described by T. The *selection type* $\oplus \langle \mathsf{p}, \{l_i : T_i\}_{i \in I} \rangle$ represents the transmission to participant p of a label l_i chosen in the set $\{l_i \mid i \in I\}$ followed by the communications described by T_i. The *receive* and *branching* types are dual of send and selection types. Recursion is guarded also in session types, and we consider them modulo fold/unfold as done for global types.

The relation between global and session types is formalised by the notion of projection as in [43].

Definition 1. *The* projection of a global type G onto a participant q ($G \upharpoonright \mathsf{q}$) *is defined by induction on G:*

$$(p \to p' : \langle U \rangle.G') \restriction q = \begin{cases} !\langle p', U \rangle (G' \restriction q) & \text{if } q = p \\ ?(p, U)(G' \restriction q) & \text{if } q = p' \\ G' \restriction q & \text{otherwise} \end{cases}$$

$$(p \to p' : \{l_i : G_i\}_{i \in I}) \restriction q = \begin{cases} \oplus \langle p', \{l_i : T_i\}_{i \in I} \rangle & \text{if } q = p \\ \&(p, \{l_i : G_i \restriction q\}_{i \in I}) & \text{if } q = p' \\ G_{i_0} \restriction q & \text{where } i_0 \in I \text{ if } q \neq p, q \neq p' \\ & \text{and } G_i \restriction q = G_j \restriction q \text{ for all } i, j \in I. \end{cases}$$

$$(\mu t G) \restriction q = \begin{cases} \mu t (G \restriction q) & \text{if } G \restriction q \neq t \\ \text{end} & \text{otherwise} \end{cases} \qquad t \restriction q = t \qquad \text{end} \restriction q = \text{end}$$

As an example, we list two of the projections of the global types G_a and G_b of the three-buyer protocol in Sect. 2.

$$G_a \restriction 3 = ?(2, \text{string}).!\langle \{1, 2\}, \text{int} \rangle; \&(1, \{\text{ok} : ?(1, \text{string}).!\langle 1, \text{date} \rangle.\text{end}, \text{quit} : \text{end}\})$$
$$G_b \restriction 1 = ?(2, \text{int}).?(2, \mathsf{T}). \oplus \langle 2, \{\text{ok} : \text{end}, \text{quit} : \text{end}\} \rangle$$

where T is defined at page 4.

Hereafter we assume all global types are *well formed*, i.e. $G \restriction q$ is defined for all q which occur in G.

4.2 Typing Rules for Pure Processes

The typing judgements for expressions and pure processes are of the shapes:

$$\Gamma \vdash e : S \quad \text{and} \quad \Gamma \vdash P \triangleright \Delta$$

where

- Γ is the *standard environment* which associates variables to sort types, service names to closed global types and process variables to pairs of sort types and session types;
- Δ is the *session environment* which associates channels to session types.

Formally we define:

$$\Gamma ::= \emptyset \mid \Gamma, x : S \mid \Gamma, a : G \mid \Gamma, X : S\,T \quad \text{and} \quad \Delta ::= \emptyset \mid \Delta, c : T$$

assuming that we can write $\Gamma, x : S$ only if $x \notin dom(\Gamma)$, where $dom(\Gamma)$ denotes the domain of Γ, i.e., the set of identifiers which occur in Γ. We use the same convention for $a : G$, $X : S\,T$ and $c : T$ (thus we can write Δ, Δ' only if $dom(\Delta) \cap dom(\Delta') = \emptyset$).

Table 6 presents the typing rules for expressions and pure processes.

Rule (NAME) is standard: recall that u stands for x and a and S includes G.

Rule (MCAST) permits to type a request on a service identified by u, if the type of y is the p-th projection of the global type G of u and the maximum participant in G (denoted by $mp(G)$) is p. Rule (MACC) permits to type the p-th participant identified by u, which uses the channel y, if the type of y is the p-th projection of the global type G of u and $p < mp(G)$.

In the typing of the example of the three-buyer protocol the types of the channels y in Seller and z in Carol are respectively the third projection of

Table 6. Typing rules for expressions and pure processes.

$$\Gamma, u : S \vdash u : S \;\; (\textsc{Name}) \qquad \Gamma \vdash \mathsf{true}, \mathsf{false} : \mathsf{bool} \;\; (\textsc{Bool}) \qquad \frac{\Gamma \vdash e_i : \mathsf{bool} \qquad (i = 1, 2)}{\Gamma \vdash e_1 \text{ and } e_2 : \mathsf{bool}} \;\; (\textsc{And})$$

$$\frac{\Gamma \vdash u : \mathsf{G} \quad \Gamma \vdash P \triangleright \Delta, y : \mathsf{G} {\upharpoonright} \mathsf{p} \quad \mathsf{p} = mp(\mathsf{G})}{\Gamma \vdash \overline{u}[\mathsf{p}](y).P \triangleright \Delta} \;\; (\textsc{MCast})$$

$$\frac{\Gamma \vdash u : \mathsf{G} \quad \Gamma \vdash P \triangleright \Delta, y : \mathsf{G} {\upharpoonright} \mathsf{p} \quad \mathsf{p} < mp(\mathsf{G})}{\Gamma \vdash u[\mathsf{p}](y).P \triangleright \Delta} \;\; (\textsc{MAcc})$$

$$\frac{\Gamma \vdash e : S \quad \Gamma \vdash P \triangleright \Delta, c : T}{\Gamma \vdash c!\langle \mathsf{p}, e \rangle.P \triangleright \Delta, c :\, !\langle \mathsf{p}, S \rangle.T} \;\; (\textsc{Send}) \qquad \frac{\Gamma, x : S \vdash P \triangleright \Delta, c : T}{\Gamma \vdash c?(\mathsf{q}, x).P \triangleright \Delta, c :\, ?(\mathsf{q}, S).T} \;\; (\textsc{Rcv})$$

$$\frac{\Gamma \vdash P \triangleright \Delta, c : T}{\Gamma \vdash c!\langle\!\langle \mathsf{p}, c' \rangle\!\rangle.P \triangleright \Delta, c :\, !\langle \mathsf{p}, T \rangle.T, c' : T} \;\; (\textsc{Deleg}) \qquad \frac{\Gamma \vdash P \triangleright \Delta, c : T, y : T}{\Gamma \vdash c?((\mathsf{q}, y)).P \triangleright \Delta, c :\, ?(\mathsf{q}, T).T} \;\; (\textsc{SRcv})$$

$$\frac{\Gamma \vdash P \triangleright \Delta, c : T_j \quad j \in I}{\Gamma \vdash c \oplus \langle \mathsf{p}, l_j \rangle.P \triangleright \Delta, c :\, \oplus\langle \mathsf{p}, \{l_i : T_i\}_{i \in I} \rangle} \;\; (\textsc{Sel})$$

$$\frac{\Gamma \vdash P_i \triangleright \Delta, c : T_i \quad \forall i \in I}{\Gamma \vdash c \&(\mathsf{p}, \{l_i : P_i\}_{i \in I}) \triangleright \Delta, c :\, \&(\mathsf{p}, \{l_i : T_i\}_{i \in I})} \;\; (\textsc{Branch})$$

$$\frac{\Gamma \vdash P \triangleright \Delta \quad \Gamma \vdash Q \triangleright \Delta'}{\Gamma \vdash P \mid Q \triangleright \Delta, \Delta'} \;\; (\textsc{Par}) \qquad \frac{\Gamma \vdash e : \mathsf{bool} \quad \Gamma \vdash P \triangleright \Delta \quad \Gamma \vdash Q \triangleright \Delta}{\Gamma \vdash \text{if } e \text{ then } P \text{ else } Q \triangleright \Delta} \;\; (\textsc{If})$$

$$\frac{\Delta \text{ end only}}{\Gamma \vdash \mathbf{0} \triangleright \Delta} \;\; (\textsc{Inact}) \qquad \frac{\Gamma, a : \mathsf{G} \vdash P \triangleright \Delta}{\Gamma \vdash (\nu a)P \triangleright \Delta} \;\; (\textsc{NRes})$$

$$\frac{\Gamma \vdash e : S \quad \Delta \text{ end only}}{\Gamma, X : S\, T \vdash X\langle e, c \rangle \triangleright \Delta, c : T} \;\; (\textsc{Var}) \qquad \frac{\Gamma, X : S\, \mathsf{t}, x : S \vdash P \triangleright y : T \quad \Gamma, X : S\, \mu\mathsf{t}.T \vdash Q \triangleright \Delta}{\Gamma \vdash \mathsf{def}\, X(x, y) = P \text{ in } Q \triangleright \Delta} \;\; (\textsc{Def})$$

G_a and the first projection of G_b. By applying rule (MCast) we can then derive $a : \mathsf{G}_a \vdash \mathsf{Seller} \triangleright \emptyset$. Similarly by applying rule (MAcc) we can derive $b : \mathsf{G}_b \vdash \mathsf{Carol} \triangleright \emptyset$. (The processes Seller and Carol are defined in Table 1.)

The successive six rules associate the input/output processes to the input/output types in the expected way. For example we can derive:

$$\vdash t \oplus \langle \{2, 3\}, \mathsf{ok} \rangle.t!\langle 3, \texttt{"Address"} \rangle; t?(3, date).\mathbf{0} \triangleright \{t : \mathsf{T}\}$$

where $\mathsf{T} = \oplus\langle \{2, 3\}, \{\mathsf{ok} :\, !\langle 3, \mathsf{string} \rangle.?(3, date).\mathsf{end}, \;\; \mathsf{quit} :\; \mathsf{end}\}\rangle$. Note that, according to our notational convention on environments, in rule (Deleg) the channel which is sent cannot appear in the session environment of the premise, i.e., $c' \notin dom(\Delta) \cup \{c\}$.

Rule (PAR) permits to put in parallel two processes only if their session environments have disjoint domains.

In rules (INACT) and (VAR) we take environments Δ which associate end to arbitrary channels, denoted by "Δ end only".

The present formulation of rule (DEF) forces to type process variables only with μ-types, while the formulation in [3,43]:

$$\frac{\Gamma, X : S\,T, x : S \vdash P \rhd y : T \qquad \Gamma, X : S\,T \vdash Q \rhd \Delta}{\Gamma \vdash \mathsf{def}\ X(x,y) = P\ \mathsf{in}\ Q \rhd \Delta}$$

allows to type unguarded process variables with arbitrary types, which can be meaningless. For example with the more permissive rule we can derive

$$\vdash \mathsf{def}\ X(x,y) = X(x,y)\ \mathsf{in}\ X\langle\mathsf{true}, z\rangle \rhd \{z : \mathsf{T}\}$$

for an arbitrary closed T, while in our system we cannot type this process since its only possible type would be $\mu\mathbf{t}.\mathbf{t}$, which is not guarded and then forbidden.

4.3 Types and Typing Rules for Runtime Processes

In this subsection we extend the communication type system to processes containing queues. We start by defining the types of queues.

$$
\begin{aligned}
\text{Message Types } M ::= \ & !\langle\mathsf{p}, U\rangle && \textit{message send} \\
| \ & \oplus\langle\mathsf{p}, l\rangle && \textit{message selection} \\
| \ & M; M && \textit{message sequence}
\end{aligned}
$$

$$
\begin{aligned}
\text{Generalised} \qquad \tau ::= \ & \mathsf{T} && \textit{session} \\
| \ & \mathsf{M} && \textit{message} \\
| \ & M; T && \textit{continuation}
\end{aligned}
$$

Message types are the types for queues: they represent the messages contained in the queues. The *message send type* $!\langle\mathsf{p}, U\rangle$ expresses the presence in a queue of an element of type U to be communicated to participant p. The *message selection type* $\oplus\langle\mathsf{p}, l\rangle$ represents the communication to participant p of the label l and $M; M$ represents sequencing of message types (we assume associativity for ";"). For example $\oplus\langle\{1,3\}, \mathsf{ok}\rangle$ is the message type for the message $(2, \{1,3\}, \mathsf{ok})$.

A *generalised type* is either a session type, or a message type, or a message type followed by a session type. Type $M; T$ represents the continuation of the type M associated to a queue with the type T associated to a pure process. Examples of generalised types are

$$!\langle 3, \mathsf{string}\rangle.?(3, \mathsf{date}).\mathsf{end} \quad \text{and} \quad !\langle 3, \mathsf{string}\rangle; ?(3, \mathsf{date}).\mathsf{end},$$

which only differ for the replacement of the leftmost "." by ";". In the first the type $!\langle 3, \mathsf{string}\rangle$ corresponds to an output action sending a string to participant 3, while in the second type $!\langle 3, \mathsf{string}\rangle$ corresponds to a message for participant

3 with a value of type string. See the examples of typing judgements at the end of this subsection.

In the typing rules for single queues the turnstile $\vdash$ is decorated with $\{s\}$ (where s is the session name of the current queue) and the session environments are mappings from channels to message types. The empty queue has the empty session environment. Each message adds an output type to the current type of the channel which has the role of the message sender. Table 7 lists the typing rules for queues, where all types in session environments are message types. The operator ";" between an arbitrary session environment and a session environment containing only one association is defined by:

$$\Delta; \{s[\mathsf{q}] : M\} = \begin{cases} \Delta', s[\mathsf{q}] : M'; M & \text{if } \Delta = \Delta', s[\mathsf{q}] : M', \\ \Delta, s[\mathsf{q}] : M & \text{otherwise.} \end{cases}$$

For example we can derive $\vdash_{\{s\}} s : (3, \{1, 2\}, \mathsf{ok}) \rhd \{s[3] : \oplus\langle\{1, 2\}, \mathsf{ok}\rangle\}$.

Table 7. Typing rules for queues.

$$\frac{}{\Gamma \vdash_{\{s\}} s : \varnothing \rhd \boldsymbol{0}} \text{ (QINIT)} \qquad \frac{\Gamma \vdash_{\{s\}} s : h \rhd \Delta \qquad \Gamma \vdash v : S}{\Gamma \vdash_{\{s\}} s : h \cdot (\mathsf{q}, \mathsf{p}, v) \rhd \Delta; \{s[\mathsf{q}] : !\langle\mathsf{p}, S\rangle\}} \text{ (QSEND)}$$

$$\frac{\Gamma \vdash_{\{s\}} s : h \rhd \Delta}{\Gamma \vdash_{\{s\}} s : h \cdot (\mathsf{q}, \mathsf{p}, s'[\mathsf{p}']) \rhd (\Delta; \{s[\mathsf{q}] : !\langle\mathsf{p}, \mathsf{T}\rangle\}), s'[\mathsf{p}'] : \mathsf{T}} \text{ (QDELEG)}$$

$$\frac{\Gamma \vdash_{\{s\}} s : h \rhd \Delta}{\Gamma \vdash_{\{s\}} s : h \cdot (\mathsf{q}, \mathsf{p}, l) \rhd \Delta; \{s[\mathsf{q}] : \oplus\langle\mathsf{p}, l\rangle\}} \text{ (QSEL)}$$

For typing pure processes in parallel with queues, we need to use generalised types in session environments and to add further typing rules.

In order to take into account the structural congruence between queues (see Table 3) we consider message types modulo the equivalence relation $\approx$ induced by the rule:

$$M; \natural\langle\mathsf{p}, Z\rangle; \natural'\langle\mathsf{p}', Z\rangle; M' \approx M; \natural'\langle\mathsf{p}', Z\rangle; \natural\langle\mathsf{p}, Z\rangle; M' \quad \text{if } \mathsf{p} \neq \mathsf{p}'$$

where $\natural \in \{!, \oplus\}$ and $Z \in \{U, l\}$).

The equivalence relation on message types extends to generalised types by:

$$M \approx M' \text{ implies } M; \tau \approx M'; \tau$$

We say that two session environments Δ and Δ' are equivalent (notation $\Delta \approx \Delta'$) if $c : \tau \in \Delta$ and $\tau \neq \mathsf{end}$ imply $c : \tau' \in \Delta'$ with $\tau \approx \tau'$ and vice versa. The reason for ignoring end types is that rules (INACT) and (VAR) allow to freely introduce them.

In composing two session environments we want to put in sequence a message type and a session type for the same channel with role. For this reason we define the partial composition $*$ between generalised types as:

$$\tau * \tau' = \begin{cases} \tau ; \tau' & \text{if } \tau \text{ is a message type,} \\ \tau' ; \tau & \text{if } \tau' \text{ is a message type.} \end{cases}$$

Notice that $\tau * \tau'$ is defined only if at least one between τ and τ' is a message type.

We extend $*$ to session environments as expected:

$$\Delta * \Delta' = \Delta \backslash dom(\Delta') \cup \Delta' \backslash dom(\Delta) \cup \{c : \tau * \tau' \mid c : \tau \in \Delta \ \wedge \ c : \tau' \in \Delta'\}.$$

Note that $*$ is commutative, i.e., $\Delta * \Delta' = \Delta' * \Delta$. Also if we can derive message types only for channels with roles, we consider channel variables in the definition of $*$ for session environments, since we want to get for example that $\{y : \mathsf{end}\} * \{y : \mathsf{end}\}$ is undefined (message types do not contain end).

To give the rules for typing processes with queues we introduce consistency of session environments, which assures that each pair of participants in a multiparty conversation performs their mutual communications in a consistent way. Consistency is defined using the notions of projection of generalised types and of duality, given respectively in Definitions 2 and 3. Notice that projection is not defined for message types.

Definition 2. *The* partial *projection of the generalised type* τ *onto* q, *denoted by* $\tau \upharpoonright \mathsf{q}$, *is defined by:*

$$(!\langle \mathsf{p}, U \rangle.T) \upharpoonright \mathsf{q} = \begin{cases} !U.T \upharpoonright \mathsf{q} & \text{if } \mathsf{q} = \mathsf{p}, \\ T \upharpoonright \mathsf{q} & \text{otherwise.} \end{cases} \qquad (?\langle \mathsf{p}, U \rangle.T) \upharpoonright \mathsf{q} = \begin{cases} ?U.T \upharpoonright \mathsf{q} & \text{if } \mathsf{p} = \mathsf{q}, \\ T \upharpoonright \mathsf{q} & \text{otherwise.} \end{cases}$$

$$(\,!\langle \mathsf{p}, U \rangle; \tau') \upharpoonright \mathsf{q} = \begin{cases} !U; \tau' \upharpoonright \mathsf{q} & \text{if } \mathsf{q} = \mathsf{p}, \\ \tau' \upharpoonright \mathsf{q} & \text{otherwise.} \end{cases} \qquad (\oplus \langle \mathsf{p}, l \rangle; \tau') \upharpoonright \mathsf{q} = \begin{cases} \oplus l; \tau' \upharpoonright \mathsf{q} & \text{if } \mathsf{q} = \mathsf{p}, \\ \tau' \upharpoonright \mathsf{q} & \text{otherwise.} \end{cases}$$

$$(\oplus \langle \mathsf{p}, \{l_i : T_i\}_{i \in I} \rangle) \upharpoonright \mathsf{q} = \begin{cases} \oplus \{l_i : T_i \upharpoonright \mathsf{q}\}_{i \in I} & \text{if } \mathsf{q} = \mathsf{p}, \\ T_{i_0} \upharpoonright \mathsf{q} & \text{where } i_0 \in I \text{ if } \mathsf{q} \neq \mathsf{p} \text{ and } T_i \upharpoonright \mathsf{q} = T_j \upharpoonright \mathsf{q} \text{ for all } i, j \in I. \end{cases}$$

$$(\&\langle \mathsf{p}, \{l_i : T_i\}_{i \in I} \rangle) \upharpoonright \mathsf{q} = \begin{cases} \&\{l_i : T_i \upharpoonright \mathsf{q}\}_{i \in I} & \text{if } \mathsf{q} = \mathsf{p}, \\ T_{i_0} \upharpoonright \mathsf{q} & \text{where } i_0 \in I \text{ if } \mathsf{q} \neq \mathsf{p} \text{ and } T_i \upharpoonright \mathsf{q} = T_j \upharpoonright \mathsf{q} \text{ for all } i, j \in I. \end{cases}$$

$$(\mu t.T) \upharpoonright \mathsf{q} = \begin{cases} \mu t.(T \upharpoonright \mathsf{q}) & \text{if } T \upharpoonright \mathsf{q} \neq t, \\ \mathsf{end} & \text{otherwise.} \end{cases} \qquad t \upharpoonright \mathsf{q} = t \qquad \mathsf{end} \upharpoonright \mathsf{q} = \mathsf{end}$$

Definition 3. *The* duality relation *between projections of generalised types* $(\bowtie)$ *is the minimal symmetric relation which satisfies:*

$$\mathsf{end} \bowtie \mathsf{end} \qquad t \bowtie t \qquad \mathfrak{T} \bowtie \mathfrak{T}' \implies \mu t.\mathfrak{T} \bowtie \mu t.\mathfrak{T}'$$

$$\mathfrak{T} \bowtie \mathfrak{T}' \implies !U.\mathfrak{T} \bowtie ?U.\mathfrak{T}' \qquad \mathfrak{T} \bowtie \mathfrak{T}' \implies !U; \mathfrak{T} \bowtie ?U.\mathfrak{T}'$$

$$\forall i \in I \ \mathfrak{T}_i \bowtie \mathfrak{T}_i' \implies \oplus \{l_i : \mathfrak{T}_i\}_{i \in I} \bowtie \&\{l_i : \mathfrak{T}_i'\}_{i \in I}$$

$$\exists i \in I \ l = l_i \ \wedge \ \mathfrak{T} \bowtie \mathfrak{T}_i \implies \oplus l; \mathfrak{T} \bowtie \&\{l_i : \mathfrak{T}_i\}_{i \in I}$$

where $\mathfrak{T}$ *ranges over projections of generalised types.*

Definition 4. *A session environment* Δ *is* consistent *for the session* s *(notation* $\mathsf{co}(\Delta, s)$*) if* $s[\mathsf{p}] : \tau \in \Delta$ *and* $s[\mathsf{q}] : \tau' \in \Delta$ *imply* $\tau \upharpoonright \mathsf{q} \bowtie \tau' \upharpoonright \mathsf{p}$. *A session environment is* consistent *if it is consistent for all sessions which occur in it.*

It is easy to check that projections of a same global type are always dual.

Proposition 1. *Let G be a global type and $\mathsf{p} \neq \mathsf{q}$. Then $(G \upharpoonright \mathsf{p}) \upharpoonright \mathsf{q} \bowtie (G \upharpoonright \mathsf{q}) \upharpoonright \mathsf{p}$.*

This proposition assures that session environments obtained by projecting global types are always consistent.

The vice versa is not true, i.e. there are consistent session environments which are not projections of global types. An example is:

$$\{s[1] :?(2, \mathsf{bool}).!(3, \mathsf{bool}).\mathsf{end}, \, s[2] :?(3, \mathsf{bool}).!(1, \mathsf{bool}).\mathsf{end}, \, s[3] :?(1, \mathsf{bool}).!(2, \mathsf{bool}).\mathsf{end}\}$$

Note that for sessions with only two participants, instead, all consistent session environments are projections of global types.

Table 8. Typing rules for processes.

$$\frac{\Gamma \vdash P \triangleright \Delta}{\Gamma \vdash_\emptyset P \triangleright \Delta} \; (\textsc{GInit}) \qquad \frac{\Gamma \vdash_\Sigma P \triangleright \Delta \quad \Delta \approx \Delta'}{\Gamma \vdash_\Sigma P \triangleright \Delta'} \; (\textsc{Equiv})$$

$$\frac{\Gamma \vdash_\Sigma P \triangleright \Delta \quad \Gamma \vdash_{\Sigma'} Q \triangleright \Delta' \quad \Sigma \cap \Sigma' = \emptyset}{\Gamma \vdash_{\Sigma \cup \Sigma'} P \mid Q \triangleright \Delta * \Delta'} \; (\textsc{GPar})$$

$$\frac{\Gamma \vdash_\Sigma P \triangleright \Delta \quad \mathsf{co}(\Delta, s)}{\Gamma \vdash_{\Sigma \backslash s} (\nu s) P \triangleright \Delta \backslash s} \; (\textsc{GSRes}) \qquad \frac{\Gamma, a : G \vdash_\Sigma P \triangleright \Delta}{\Gamma \vdash_\Sigma (\nu a) P \triangleright \Delta} \; (\textsc{GNRes})$$

$$\frac{\Gamma, X : S\,\mathsf{t}, x : S \vdash P \triangleright \{y : T\} \quad \Gamma, X : S\,\mu\mathsf{t}.T \vdash_\Sigma Q \triangleright \Delta}{\Gamma \vdash_\Sigma \mathsf{def}\ X(x, y) = P\ \mathsf{in}\ Q \triangleright \Delta} \; (\textsc{GDef})$$

Table 8 lists the typing rules for processes containing queues. The judgement

$$\Gamma \vdash_\Sigma P \triangleright \Delta$$

means that P contains the queues whose session names are in Σ. Rule (GINIT) promotes the typing of a pure process to the typing of an arbitrary process without session names, since a pure process does not contain queues. When two arbitrary processes are put in parallel (rule (GPAR)) we need to require that each session name is associated to at most one queue (condition $\Sigma \cap \Sigma' = \emptyset$).

Examples of derivable judgements are:

$$\vdash_{\{s\}} P \mid s : (3, \{1, 2\}, \mathsf{ok}) \triangleright \{s[3] : \oplus\langle\{1, 2\}, \mathsf{ok}\rangle; !\langle 1, \mathsf{string}\rangle.?(1, \mathsf{date}).\mathsf{end}\}$$

where $P = s[3]!\langle 1, \texttt{"Address"}\rangle; s[3]?(1, date); \mathbf{0}$ and

$$\vdash_{\{s\}} P' \mid s : (3, \{1, 2\}, \mathsf{ok}) \cdot (3, 1, \texttt{"Address"}) \triangleright \{s[3] : \oplus\langle\{1, 2\}, \mathsf{ok}\rangle; !\langle 1, \mathsf{string}\rangle; ?(1, \mathsf{date}).\mathsf{end}\}$$

where $P' = s[3]?(1, \mathsf{date}); \mathbf{0}$. Note that

$$P \mid s : (3, \{1, 2\}, \mathsf{ok}) \longrightarrow P' \mid s : (3, \{1, 2\}, \mathsf{ok}) \cdot (3, 1, \mathsf{string})$$

A simple example showing that consistency is necessary for subject reduction is the process:

$$P = s[1]!\langle 2, \mathsf{true}\rangle.s[1]?(2, x).\mathbf{0} \mid s[2]?(1, x').s[2]!\langle 1, x' + 1\rangle.\mathbf{0}$$

which can be typed with the non consistent session environment

$$\{s[1] :!\langle 2, \mathsf{bool}\rangle.?(2, \mathsf{nat}).\mathsf{end}, s[2] :?(1, \mathsf{nat}).!\langle 1, \mathsf{nat}\rangle.\mathsf{end}\}$$

In fact P reduces to the process

$$s[1]?(2, x).\mathbf{0} \mid s[2]!\langle 1, \mathsf{true} + 1\rangle.\mathbf{0}$$

which cannot be typed and it is stuck.

4.4   Subject Reduction

Since session environments represent the forthcoming communications, by reducing processes session environments can change. This can be formalised as in [43] by introducing the notion of reduction of session environments, whose rules are:

- $\{s[\mathbf{p}] : M; !\langle \mathbf{q}, U\rangle.T\} \Rightarrow \{s[\mathbf{p}] : M; !\langle \mathbf{q}, U\rangle; T\}$
- $\{s[\mathbf{p}] : !\langle \mathbf{q}, U\rangle; \tau, s[\mathbf{q}] : M; ?(\mathbf{p}, U).T\} \Rightarrow \{s[\mathbf{p}] : \tau, s[\mathbf{q}] : M; T\}$
- $\{s[\mathbf{p}] : M; \oplus\langle \mathbf{p}, \{l_i : T_i\}_{i\in I}\rangle\} \Rightarrow \{s[\mathbf{p}] : M; \oplus(\mathbf{p}, l_j); T_j\}$ for $j \in I$
- $\{s[\mathbf{p}] : \oplus\langle \mathbf{q}, l\rangle; \tau, s[\mathbf{q}] : M; \&(\mathbf{p}, \{l_i : T_i\}_{i\in I})\} \Rightarrow \{s[\mathbf{p}] : \tau, s[\mathbf{q}] : M; T_i\}$ if $l = l_i$
- $\Delta, \Delta'' \Rightarrow \Delta', \Delta''$ if $\Delta \Rightarrow \Delta'$

where M can be missing and message types are considered modulo the equivalence relation $\approx$ defined at page 14.

The first rule corresponds to putting in a queue a message with sender $\mathbf{p}$, receiver $\mathbf{q}$ and content of type U. The second rule corresponds to reading from a queue a message with sender $\mathbf{p}$, receiver $\mathbf{q}$ and content of type U. The third and fourth rules are similar, but a label is transmitted.

Notice that not all the left-hand-sides of the reduction rules for processes are typed by consistent session environments. For example,

$$\Gamma \vdash_{\Sigma} s[1]?(2, x).s[1]?(2, y).\mathbf{0} \mid s : (2, 1, \mathsf{true}) \rhd \{s[1] :?(2, \mathsf{bool}).?(2, \mathsf{nat}).\mathsf{end}, s : [2] : !\langle \mathsf{bool}, 1\rangle\}$$

Observe that $s[1]?(2, x).s[1]?(2, y).\mathbf{0} \mid s : (2, 1, \mathsf{true})$ matches the left-hand-side of the reduction rule [Rcv] and $\{s[1] :?(2, \mathsf{bool}).?(2, \mathsf{nat}).\mathsf{end}, s : [2] : !\langle \mathsf{bool}, 1\rangle\}$ is not consistent. The process obtained by putting this network in parallel with $s[2]!\langle 1, 7\rangle.\mathbf{0}$ has a consistent session environment. It is then crucial to show that if the left-hand-side of a reduction rule is typed by a session environment, which is consistent when composed with some other session environment, then the same property holds for the right-hand-side too. It is sufficient to consider the reduction rules which do not contain process reductions as premises, i.e. which are the leaves in the reduction trees. This is formalised in the following lemma, which is the key step for proving the Subject Reduction Theorem.

Lemma 1 (Main Lemma). *Let $\Gamma \vdash_\Sigma P \triangleright \Delta$, and $P \longrightarrow P'$ be obtained by any reduction rule different from [Ctxt], [Str], and $\Delta * \Delta_0$ be consistent, for some Δ_0. Then there is Δ' such that $\Gamma \vdash_\Sigma P' \triangleright \Delta'$ and $\Delta \Rightarrow^* \Delta'$ and $\Delta' * \Delta_0$ is consistent.*

We end this section by formulating subject reduction.

Theorem 1 (Subject Reduction). *If $\Gamma \vdash_\Sigma P \triangleright \Delta$ with Δ consistent and $P \longrightarrow^* P'$, then $\Gamma \vdash_\Sigma P' \triangleright \Delta'$ for some consistent Δ' such that $\Delta \Rightarrow^* \Delta'$.*

Appendix A proves subject reduction. Note that communication safety and protocol fidelity easily follow from Theorem 1.

5   Related Work

5.1   Multiparty Session Types

The first theoretical works on multiparty session types are [10, 43]. The paper [10] uses a distributed calculus where each channel connects a master endpoint to one or more slave endpoints; instead of global types, they solely use (recursion-free) local types. For type checking, local types are projected to binary sessions, so that type safety is ensured using duality, but it loses sequencing information: hence progress in a session interleaved with other sessions is not guaranteed.

In this article we have presented the calculus of [25], which is an essential improvement and simplification of the calculus in [43]. Both processes and types in [43] share a vector of channels and each communication uses one of these channels. In the present work, instead, processes and types use indexes for identifying the participants of a session.

The communication type system in this article improves the one of [43] in two main technical points without sacrificing expressiveness. First, it avoids the overhead of global linearity-check in [43] because our global types automatically satisfy the linearity condition in [43] due to the limitation to bi-directional channel communications. Second, it provides a more liberal policy in the use of variables in delegation, since we do not require to delegate a set of session channels. The global types in [43] have a parallel composition operator, but its projectability from global to local types limits to disjoint senders and receivers; hence our global types do not affect the expressivity.

5.2   Theoretical Studies on Multiparty Session Types

Extensions of the original multiparty session types [43] and of the communication type system in this article have been proposed, often motivated by use cases resulting from industry applications (Sect. 5.8). Such extensions include: a subtyping for asynchronous multiparty session types enhancing efficiency [52], motivated by financial protocols and multicore algorithms; parametrised global types for parallel programming and Web service descriptions [34]; communication buffered analysis [30]; extensions to the sumtype and its encoding [61] for describing Healthcare workflows; exception handling for multiparty conversations [15] for Web services and financial protocols; a liveness-preserving refinement for multiparty session types [64].

Multiparty session types can be extended with logical assertions following the design by contract framework [7]. This framework is enriched in [6] to handle stateful logical assertions, while [21] offers more fine-grained property analysis for multiparty session types with these stateful assertions.

In [31] roles are inhabited by an arbitrary number of participants which can dynamically join and leave a session. The paper [71] shows that the multirole session types [31] can be naturally represented in a dependent-typed language.

To enhance expressivity and flexibility of multiparty session types, the work [28] proposes nested, higher-order multiparty session types and the work [18] studies a generalisation of choices and parallelism. The paper [17] directly types a global description language [16] by multiparty session types without using local types. This direct approach can type processes which are untypable in the original multiparty session typing (i.e. the communication type system in this article). The paper [51] extends the work in [17] to compositional global description languages.

As another line of the study, we extend the multiparty session types to express temporal properties [9]. In this work, the global times are enriched with time constraints, in a way similar to timed automata.

A type system enforcing a stronger correspondence between nondeterministic choices expressed in multiparty session types and the behaviour of processes involved in multiparty sessions has been investigated in [8].

An overview of the recent developments in these studies is the survey in the state-of-the art report produced by the Foundations Working Group of the IC COST Action BETTY, entitled "Foundations of Behavioural Types" [45].

5.3   Progress and Session Interleaving

Multiparty session types are a convenient methodology for ensuring progress of systems of communicating processes. However, progress is only guaranteed within a *single* session [31,35,43], but not when multiple sessions are interleaved. The first papers considering progress for interleaved sessions required the nesting of sessions in Java [24,36]. These systems can guarantee progress for only one single active binary session. The work [25] develops a static interaction type system for global progress in dynamically interleaved and interfered multiparty sessions. A type inference algorithm for this system has been studied in [22], although for finite types only. The work [63, technical report] presents a type system for the linear π-calculus that can ensure progress even in presence of session interleaving, exploiting an encoding similar to that described in [27] of sessions into the linear π-calculus. However, not *all* multiparty sessions can be encoded into well-typed linear π-calculus processes. In this respect, the richer structure of multiparty session types increases the range of systems for which non-trivial properties such as progress can be guaranteed.

5.4 Security

Enforcement of *integrity* properties in multiparty sessions, using session types, has been studied in [4,67]. These papers propose a compiler which, given a multiparty session description, implements cryptographic protocols that guarantee session execution integrity.

The work [14] and in its extended version [12] propose a session type system for a calculus of multiparty sessions enriched with security levels, adding access control and secure information flow requirements in the typing rules, and show that this type system guarantees preservation of data confidentiality during session execution. In [13] this calculus is equipped with a monitored semantics, which blocks the execution of processes as soon as they attempt to leak information, raising an error.

Various approaches for enforcing security into calculi and languages for structured communications have been recently surveyed in the state-of-the art report produced by the Security Working Group of the IC COST Action BETTY, entitled "Combining Behavioural Types with Security Analysis" [2].

5.5 Behavioural Semantics

Typed behavioural theory has been one of the central topics in the study of the π-calculus throughout its history, for example, reasoning about various encodings into the typed π-calculi [47,65,74]. In the context of typed bisimulations and reduction-closed theories, the work [46] shows that unique behavioural theories can be constructed based on the multiparty session types. The behavioural theory in [46] treats the mutual effects of multiple choreographic sessions which are shared among distributed participants as their common knowledge or agreements, reflecting the origin of choreographic frameworks [73]. These features related to multiparty session type discipline make the theory distinct from any type-based bisimulations in the literature and also applicable to a real choreographic usecase from a large-scale distributed system. This bisimulation is called *globally governed*, since it uses global multiparty specifications to regulate the conversational behaviour of distributed processes.

5.6 Runtime Monitoring and Adaptation

Multiparty session types were originally developed to be used for static type checking of communicating processes. Via collaborations with Ocean Observatories Initiative [62], it was discovered that the framework of multiparty session types can be naturally extended to runtime type checking (monitoring). A formulation of the runtime monitoring (dynamic or runtime type checking) is firstly proposed in [20]. Later the work [5] has formally proved its correctness and properties guaranteed by the runtime monitoring based on multiparty session types. See Sect. 5.8.

Works addressing adaptation for multiparty communications include [19,23, 26]. The paper [26] proposes a choreographic language for distributed applications. Adaptation follows a rule-based approach, in which all interactions, under all possible changes produced by the adaptation rules, proceed as prescribed by an abstract model. In [23] a calculus based on global types, monitors and processes is introduced and adaptation is triggered after the execution of the communications prescribed by a global type, in reaction to changes of the global state. In contrast, in [19] adaptation is triggered by security violations, and assures access control and secure information flow.

5.7 Linkages with Other Frameworks

The work [32] gives a linkage between communicating automata [11] and a general graphical version of multiparty session types, proving a correspondence between the safety properties of communicating automata and multiparty session types. The paper [33] studies the sound and complete characterisation of the multiparty session types in communicating automata and applies the result to the synthesis of the multiparty session types. The inference of global types from a set of local types is also studied in [48]. The techniques developed in [33,48] are extended to a synthesis of general graphical multiparty session types in [49].

The recent work [37] studies the relationship of multiparty session types with Petri Nets. It proposes a conformance relation between global session nets and endpoint programs, and proves its safety.

5.8 Implementations Based on Multiparty Session Types

The research group led by the last author is currently designing and implementing a modelling and specification language with multiparty session types [68,69] in collaboration with some industrial partners [40,41]. This protocol language is called Scribble. An article [75] also explains the origin and recent development on Scribble.

Java protocol optimisation [70] based on multiparty session types and generation of multiparty cryptographic protocols [4] are also studied. The multiparty session type theory is applied to Healthcare workflows [38]. Its prototype implementation (the multiparty session π-processes with sumtypes) is available from [1].

Based on the runtime type checking theory, we are implementing a runtime monitoring [29,44,55] under collaborations with Ocean Observatories Initiative [62]. The work [29,44] allows interruptions in Scribble and proves the correctness of this extension. Further we generalise the Python implementation to the Actor framework [54]. In order to express temporal properties studied in timed multiparty session types [9], the work [53] extends Scribble with timed constrains and implements the runtime monitoring in Python.

We also apply the multiparty session types to high-performance parallel programming in C [58,60] and MPI [57]. A parametrised version of Scribble [57,59] based on the theory of parametrised multiparty session types [34] is developed. This extension, called Pabble, is used for automatically generating MPI parallel programs from sequential C code in [56].

Acknowledgements. The research reported in this chapter has been partially supported by COST IC1201. The first three authors have been partially supported by MIUR PRIN Project CINA Prot. 2010LHT4KM and Torino University/Compagnia San Paolo Project SALT. The last author has been partially supported by EPSRC EP/K011715/01, EP/K034413/01 and EP/L00058X/1 and the EU project FP7-612985 UpScale.

A   Properties of the Communication Type System

This appendix completes the description of the communication type system given in Sect. 4. Auxiliary lemmas, in particular inversion lemmas, are the content of Sect. A.1. Lastly Sect. A.2 proves subject reduction.

A.1   Auxiliary Lemmas

We start with inversion lemmas which can be easily shown by induction on derivations.

Lemma 2 (Inversion Lemma for Pure Processes).

1. *If* $\Gamma \vdash u : S$, *then* $u : S \in \Gamma$.
2. *If* $\Gamma \vdash \mathsf{true} : S$, *then* $S = \mathsf{bool}$.
3. *If* $\Gamma \vdash \mathsf{false} : S$, *then* $S = \mathsf{bool}$.
4. *If* $\Gamma \vdash e_1$ *and* $e_2 : S$, *then* $\Gamma \vdash e_1 : \mathsf{bool}$ *and* $\Gamma \vdash e_2 : \mathsf{bool}$ *and* $S = \mathsf{bool}$.
5. *If* $\Gamma \vdash \overline{a}[\mathsf{p}](y).P \triangleright \Delta$, *then* $\Gamma \vdash a : \mathsf{G}$ *and* $\Gamma \vdash P \triangleright \Delta, y{:}\mathsf{G} \upharpoonright \mathsf{p}$ *and* $\mathsf{p} = mp(\mathsf{G})$.
6. *If* $\Gamma \vdash a[\mathsf{p}](y).P \triangleright \Delta$, *then* $\Gamma \vdash a : \mathsf{G}$ *and* $\Gamma \vdash P \triangleright \Delta, y{:}\mathsf{G} \upharpoonright \mathsf{p}$ *and* $\mathsf{p} < mp(\mathsf{G})$.
7. *If* $\Gamma \vdash c!\langle \mathsf{p}, e \rangle.P \triangleright \Delta$, *then* $\Delta = \Delta', c : \,!\langle \mathsf{p}, S \rangle.T$ *and* $\Gamma \vdash e : S$ *and* $\Gamma \vdash P \triangleright \Delta', c : T$.
8. *If* $\Gamma \vdash c?(\mathsf{q}, x).P \triangleright \Delta$, *then* $\Delta = \Delta', c : \,?(\mathsf{q}, S).T$ *and* $\Gamma, x : S \vdash P \triangleright \Delta', c : T$.
9. *If* $\Gamma \vdash c!\langle\!\langle \mathsf{p}, c' \rangle\!\rangle.P \triangleright \Delta$, *then* $\Delta = \Delta', c : \,!\langle \mathsf{p}, T \rangle.T, c' : T$ *and* $\Gamma \vdash P \triangleright \Delta', c : T$.
10. *If* $\Gamma \vdash c?((\mathsf{q}, y)).P \triangleright \Delta$, *then* $\Delta = \Delta', c : \,?(\mathsf{q}, T).T$ *and* $\Gamma \vdash P \triangleright \Delta', c : T, y : T$.
11. *If* $\Gamma \vdash c \oplus \langle \mathsf{p}, l_j \rangle.P \triangleright \Delta$, *then* $\Delta = \Delta', c : \oplus\langle \mathsf{p}, \{l_i : T_i\}_{i \in I} \rangle$ *and* $\Gamma \vdash P \triangleright \Delta', c : T_j$ *and* $j \in I$.
12. *If* $\Gamma \vdash c\&(\mathsf{p}, \{L_i : P_i\}_{i \in I}) \triangleright \Delta$, *then* $\Delta = \Delta', c : \&(\mathsf{p}, \{l_i : T_i\}_{i \in I})$ *and* $\Gamma \vdash P_i \triangleright \Delta', c : T_i \ \forall i \in I$.
13. *If* $\Gamma \vdash P \mid Q \triangleright \Delta$, *then* $\Delta = \Delta', \Delta''$ *and* $\Gamma \vdash P \triangleright \Delta'$ *and* $\Gamma \vdash Q \triangleright \Delta''$.
14. *If* $\Gamma \vdash \mathsf{if}\ e\ \mathsf{then}\ P\ \mathsf{else}\ Q \triangleright \Delta$, *then* $\Gamma \vdash e : \mathsf{bool}$ *and* $\Gamma \vdash P \triangleright \Delta$ *and* $\Gamma \vdash Q \triangleright \Delta$.
15. *If* $\Gamma \vdash \mathbf{0} \triangleright \Delta$, *then* Δ end *only*.
16. *If* $\Gamma \vdash (\nu a)P \triangleright \Delta$, *then* $\Gamma, a : \mathsf{G} \vdash P \triangleright \Delta$.
17. *If* $\Gamma \vdash X\langle e, c \rangle \triangleright \Delta$, *then* $\Gamma = \Gamma', X : S\ T$ *and* $\Delta = \Delta', c : T$ *and* $\Gamma \vdash e : S$ *and* Δ' end *only*.
18. *If* $\Gamma \vdash \mathsf{def}\ X(x, y) = P\ \mathsf{in}\ Q \triangleright \Delta$, *then* $\Gamma, X : S\ t, x : S \vdash P \triangleright \{y : T\}$ *and* $\Gamma, X : S\ \mu t.T \vdash Q \triangleright \Delta$.

Lemma 3 (Inversion Lemma for Processes).

1. If $\Gamma \vdash_\Sigma P \triangleright \Delta$ and P is a pure process, then $\Sigma = \emptyset$ and $\Gamma \vdash P \triangleright \Delta$.
2. If $\Gamma \vdash_\Sigma s : h \triangleright \Delta$, then $\Sigma = \{s\}$.
3. If $\Gamma \vdash_{\{s\}} s : \phi \triangleright \Delta$, then Δ endonly.
4. If $\Gamma \vdash_{\{s\}} s : h \cdot (\mathsf{q}, \mathsf{p}, v) \triangleright \Delta$, then $\Delta \approx \Delta'; \{s[\mathsf{q}] : !\langle \mathsf{p}, S \rangle\}$ and $\Gamma \vdash_{\{s\}} s : h \triangleright \Delta'$ and $\Gamma \vdash v : S$.
5. If $\Gamma \vdash_{\{s\}} s : h \cdot (\mathsf{q}, \mathsf{p}, s'[\mathsf{p'}]) \triangleright \Delta$, then $\Delta \approx (\Delta'; \{s[\mathsf{q}] : !\langle \mathsf{p}, T \rangle\}), s'[\mathsf{p'}] : T$ and $\Gamma \vdash_{\{s\}} s : h \triangleright \Delta'$.
6. If $\Gamma \vdash_{\{s\}} s : h \cdot (\mathsf{q}, \mathsf{p}, l) \triangleright \Delta$, then $\Delta \approx \Delta'; \{s[\mathsf{q}] : \oplus \langle \mathsf{p}, l \rangle\}$ and $\Gamma \vdash_{\{s\}} s : h \triangleright \Delta'$.
7. If $\Gamma \vdash_\Sigma P \mid Q \triangleright \Delta$, then $\Sigma = \Sigma_1 \cup \Sigma_2$ and $\Sigma_1 \cap \Sigma_2 = \emptyset$ and $\Delta = \Delta_1 * \Delta_2$ and $\Gamma \vdash_{\Sigma_1} P \triangleright \Delta_1$ and $\Gamma \vdash_{\Sigma_2} Q \triangleright \Delta_2$.
8. If $\Gamma \vdash_\Sigma (\nu s)P \triangleright \Delta$, then $\Sigma = \Sigma' \setminus s$ and $\Delta = \Delta' \setminus s$ and $\mathsf{co}(\Delta', s)$ and $\Gamma \vdash_{\Sigma'} P \triangleright \Delta'$.
9. If $\Gamma \vdash_\Sigma (\nu a)P \triangleright \Delta$, then $\Gamma, a : G \vdash_\Sigma P \triangleright \Delta$.
10. If $\Gamma \vdash_\Sigma \mathsf{def}\ X(x,y) = P$ in $Q \triangleright \Delta$, then $\Gamma, X : S\ t, x : S \vdash P \triangleright y : T$ and $\Gamma, X : S\ \mu t.T \vdash_\Sigma Q \triangleright \Delta$.

The following lemma allows to characterise the types due to the messages which occur in queues. The proof is standard by induction on the lengths of queues.

Lemma 4.

1. If $\Gamma \vdash_{\{s\}} s : h_1 \cdot (\mathsf{q}, \mathsf{p}, v) \cdot h_2 \triangleright \Delta$, then $\Delta = \Delta_1 * \{s[\mathsf{q}] : !\langle \mathsf{p}, S \rangle\} * \Delta_2$ and $\Gamma \vdash_{\{s\}} s : h_i \triangleright \Delta_i$ $(i = 1,2)$ and $\Gamma \vdash v : S$.
   Vice versa $\Gamma \vdash_{\{s\}} s : h_i \triangleright \Delta_i$ $(i = 1,2)$ and $\Gamma \vdash v : S$ imply

$$\Gamma \vdash_{\{s\}} s : h_1 \cdot (\mathsf{q}, \mathsf{p}, v) \cdot h_2 \triangleright \Delta_1 * \{s[\mathsf{q}] : !\langle \mathsf{p}, S \rangle\} * \Delta_2.$$

2. If $\Gamma \vdash_{\{s\}} s : h_1 \cdot (\mathsf{q}, \mathsf{p}, s'[\mathsf{p'}]) \cdot h_2 \triangleright \Delta$, then $\Delta = (\Delta_1 * \{s[\mathsf{q}] : !\langle \mathsf{p}, T \rangle\} * \Delta_2), s'[\mathsf{p'}] : T$ and $\Gamma \vdash_{\{s\}} s : h_i \triangleright \Delta_i$ $(i = 1,2)$.
   Vice versa $\Gamma \vdash_{\{s\}} s : h_i \triangleright \Delta_i$ $(i = 1,2)$ imply

$$\Gamma \vdash_{\{s\}} s : h_1 \cdot (\mathsf{q}, \mathsf{p}, s'[\mathsf{p'}]) \cdot h_2 \triangleright (\Delta_1 * \{s[\mathsf{q}] : !\langle \mathsf{p}, T \rangle\} * \Delta_2), s'[\mathsf{p'}] : T.$$

3. If $\Gamma \vdash_{\{s\}} s : h_1 \cdot (\mathsf{q}, \mathsf{p}, l) \cdot h_2 \triangleright \Delta$, then $\Delta = \Delta_1 * \{s[\mathsf{q}] : \oplus \langle \mathsf{p}, l \rangle\} * \Delta_2$ and $\Gamma \vdash_{\{s\}} s : h_i \triangleright \Delta_i$ $(i = 1,2)$.
   Vice versa $\Gamma \vdash_{\{s\}} s : h_i \triangleright \Delta_i$ $(i = 1,2)$ imply

$$\Gamma \vdash_{\{s\}} s : h_1 \cdot (\mathsf{q}, \mathsf{p}, l) \cdot h_2 \triangleright \Delta_1 * \{s[\mathsf{q}] : \oplus \langle \mathsf{p}, l \rangle\} * \Delta_2.$$

We end this subsection with two classical results: type preservation under substitution and under equivalence of processes.

Lemma 5 (Substitution Lemma).

1. *If $\Gamma, x : S \vdash P \rhd \Delta$ and $\Gamma \vdash v : S$, then $\Gamma \vdash P\{v/x\} \rhd \Delta$.*
2. *If $\Gamma \vdash P \rhd \Delta, y : T$, then $\Gamma \vdash P\{s[\mathsf{p}]/y\} \rhd \Delta, s[\mathsf{p}] : T$.*

Proof. Standard induction on type derivations, with a case analysis on the last applied rule. □

Theorem 2 (Type Preservation Under Equivalence). *If $\Gamma \vdash_\Sigma P \rhd \Delta$ and $P \equiv P'$, then $\Gamma \vdash_\Sigma P' \rhd \Delta$.*

Proof. By induction on $\equiv$. We only consider some interesting cases (the other cases are straightforward).

- $P \mid \mathbf{0} \equiv P$. First we assume $\Gamma \vdash_\Sigma P \rhd \Delta$. From $\Gamma \vdash_\emptyset \mathbf{0} \rhd \emptyset$ by applying (GPar) to these two sequents we obtain $\Gamma \vdash_\Sigma P|\mathbf{0} \rhd \Delta$.

  For the converse direction assume $\Gamma \vdash_\Sigma P|\mathbf{0} \rhd \Delta$. Using 3(7) we obtain: $\Gamma \vdash_{\Sigma_1} P \rhd \Delta_1$, $\Gamma \vdash_{\Sigma_2} \mathbf{0} \rhd \Delta_2$, where $\Delta = \Delta_1 * \Delta_2$, $\Sigma = \Sigma_1 \cup \Sigma_2$ and $\Sigma_1 \cap \Sigma_2 = \emptyset$. Using 3(1) we get $\Sigma_2 = \emptyset$, which implies $\Sigma = \Sigma_1$, and $\Gamma \vdash \mathbf{0} \rhd \Delta_2$. Using 2(15) we get Δ_2 end only which implies $\Delta_1 \approx \Delta_1 * \Delta_2$, so we conclude $\Gamma \vdash_\Sigma P \rhd \Delta_1 * \Delta_2$ by applying (Equiv).

- $P \mid Q \equiv Q \mid P$. By the symmetry of the rule we have to show only one direction. Suppose $\Gamma \vdash_\Sigma P \mid Q \rhd \Delta$. Using 3(7) we obtain $\Gamma \vdash_{\Sigma_1} P \rhd \Delta_1$, $\Gamma \vdash_{\Sigma_2} Q \rhd \Delta_2$, where $\Delta = \Delta_1 * \Delta_2$, $\Sigma = \Sigma_1 \cup \Sigma_2$ and $\Sigma_1 \cap \Sigma_2 = \emptyset$. Using (GPar) we get $\Gamma \vdash_\Sigma Q \mid P \rhd \Delta_2 * \Delta_1$. Thanks to the commutativity of $*$, we get $\Delta_2 * \Delta_1 = \Delta$ and so we are done.

- $P \mid (Q \mid R) \equiv (P \mid Q) \mid R$. Suppose $\Gamma \vdash_\Sigma P \mid (Q \mid R) \rhd \Delta$. Using 3(7) we obtain $\Gamma \vdash_{\Sigma_1} P \rhd \Delta_1$, $\Gamma \vdash_{\Sigma_2} Q \mid R \rhd \Delta_2$, where $\Delta = \Delta_1 * \Delta_2$, $\Sigma = \Sigma_1 \cup \Sigma_2$ and $\Sigma_1 \cap \Sigma_2 = \emptyset$. Using 3(7) we obtain $\Gamma \vdash_{\Sigma_{21}} Q \rhd \Delta_{21}$, $\Gamma \vdash_{\Sigma_{22}} R \rhd \Delta_{22}$ where $\Delta_2 = \Delta_{21} * \Delta_{22}$, $\Sigma_2 = \Sigma_{21} \cup \Sigma_{22}$ and $\Sigma_{21} \cap \Sigma_{22} = \emptyset$. Using (GPar) we get $\Gamma \vdash_{\Sigma_1 \cup \Sigma_{21}} P \mid Q \rhd \Delta_1 * \Delta_{21}$. Using (GPar) again we get $\Gamma \vdash_\Sigma (P \mid Q) \mid R \rhd \Delta_1 * \Delta_{21} * \Delta_{22}$ and so we are done by the associativity of $*$. The proof for the other direction is similar.

- $s : h_1 \cdot (\mathsf{q}, \mathsf{p}, v) \cdot (\mathsf{q}', \mathsf{p}', v') \cdot h_2 \equiv s : h_1 \cdot (\mathsf{q}', \mathsf{p}', v') \cdot (\mathsf{q}, \mathsf{p}, v) \cdot h_2$ where $\mathsf{p} \neq \mathsf{p}'$ or $\mathsf{q} \neq \mathsf{q}'$. We assume $\mathsf{p} \neq \mathsf{p}'$ and $\mathsf{q} = \mathsf{q}'$, the proof in the case $\mathsf{q} \neq \mathsf{q}'$ being similar and simpler. If $\Gamma \vdash_\Sigma s : h_1 \cdot (\mathsf{q}, \mathsf{p}, v) \cdot (\mathsf{q}, \mathsf{p}', v') \cdot h_2 \rhd \Delta$, then $\Sigma = \{s\}$ by Lemma 3(2). This implies $\Delta = \Delta_1 * \{s[\mathsf{q}] : \ !\langle \mathsf{p}, S \rangle; \ !\langle \mathsf{p}', S' \rangle\} * \Delta_2$ and $\Gamma \vdash_{\{s\}} s : h_i \rhd \Delta_i$ $(i = 1, 2)$ and $\Gamma \vdash v : S$ and $\Gamma \vdash v' : S'$ by Lemma 4(1). By the same lemma we can derive

$$\Gamma \vdash_{\{s\}} s : h_1 \cdot (\mathsf{q}, \mathsf{p}', v') \cdot (\mathsf{q}, \mathsf{p}, v) \cdot h_2 \rhd \Delta_1 * \{s[\mathsf{q}] : \ !\langle \mathsf{p}', S' \rangle; \ !\langle \mathsf{p}, S \rangle\} * \Delta_2,$$

and we conclude using rule (Equiv), since by definition

$$\Delta_1 * \{s[\mathsf{q}] : \ !\langle \mathsf{p}', S' \rangle; \ !\langle \mathsf{p}, S \rangle\} * \Delta_2 \approx \Delta. \qquad \square$$

A.2 Proof of Subject Reduction

We show the Main Lemma first and then the Subject Reduction Theorem.

Lemma 1 (Main Lemma). *Let* $\Gamma \vdash_\Sigma P \rhd \Delta$, *and* $P \longrightarrow P'$ *be obtained by any reduction rule different from* [Ctxt], [Str], *and* $\Delta * \Delta_0$ *be consistent, for some* Δ_0. *Then there is* Δ' *such that* $\Gamma \vdash_\Sigma P' \rhd \Delta'$ *and* $\Delta \Rightarrow^* \Delta'$ *and* $\Delta' * \Delta_0$ *is consistent.*

Proof. The proof is by cases on process reduction rules. We only consider some paradigmatic cases.

– [Init] $a[1](y).P_1 \mid ... \mid \overline{a}[n](y).P_n \longrightarrow (\nu s)(P_1\{s[1]/y_1\} \mid ... \mid P_n\{s[n]/y\} \mid s : \phi)$.
  By hypothesis $\Gamma \vdash_\Sigma a[1](y).P_1 \mid a[2](y_2).P_2 \mid \ldots \mid \overline{a}[n](y).P_n \rhd \Delta$; then, since the redex is a pure process, $\Sigma = \emptyset$ and $\Gamma \vdash a[1](y).P_1 \mid a[2](y_2).P_2 \mid \ldots \rhd$
  $\vdash \mid \overline{a}[n](y).P_n \rhd \Delta$ by Lemma 3(1). Using Lemma 2(13) on all the processes in parallel we have

$$\Gamma \vdash a[i](y).P_i \rhd \Delta_i \quad (1 \leq i \leq n-1) \tag{1}$$
$$\Gamma \vdash \overline{a}[n](y).P_n \rhd \Delta_n \tag{2}$$

where $\Delta = \bigcup_{i=1}^n \Delta_i$. Using Lemma 2(6) on (1) we have

$$\Gamma \vdash a : \mathsf{G}$$
$$\Gamma \vdash P_i \rhd \Delta_i, y{:}\mathsf{G} \restriction i \quad (1 \leq i \leq n-1). \tag{3}$$

Using Lemma 2(5) on (2) we have

$$\Gamma \vdash a : \mathsf{G}$$
$$\Gamma \vdash P_n \rhd \Delta_n, y : \mathsf{G} \restriction n \tag{4}$$

and $\mathrm{mp}(\mathsf{G}) = n$. Using Lemma 5(2) on (4) and (3) we have

$$\Gamma \vdash P_i\{s[i]/y\} \rhd \Delta_i, s[i] : \mathsf{G} \restriction i \quad (1 \leq i \leq n). \tag{5}$$

Using (PAR) on all the processes of (5) we have

$$\Gamma \vdash P_1\{s[1]/y\}|...|P_n\{s[n]/y\} \rhd \bigcup_{i=1}^n (\Delta_i, s[i] : \mathsf{G} \restriction i). \tag{6}$$

Note that $\bigcup_{i=1}^n (\Delta_i, s[i]{:}\mathsf{G} \restriction i) = \Delta, s[1]{:}\mathsf{G} \restriction 1, \ldots, s[n]{:}\mathsf{G} \restriction n$. Using (GINIT), (QINIT) and (GPAR) on (6) we derive

$$\Gamma \vdash_{\{s\}} P_1\{s[1]/y\}|...|P_n\{s[n]/y\} \mid s : \phi \rhd \Delta, s[1]{:}\mathsf{G} \restriction 1, \ldots, s[n]{:}\mathsf{G} \restriction n. \tag{7}$$

Using (GSRES) on (7) we conclude

$$\Gamma \vdash_\emptyset (\nu s)(P_1\{s[1]/y\}|...|P_n\{s[n]/y\} \mid s : \phi) \rhd \Delta$$

since $\{s[1]{:}\mathsf{G} \restriction 1, \ldots, s[n]{:}\mathsf{G} \restriction n\}$ is consistent and

$$(\Delta, s[1]{:}\mathsf{G} \restriction 1, \ldots, s[n]{:}\mathsf{G} \restriction n) \setminus s = \Delta.$$

172   M. Coppo et al.

– [Send] $s[p]!\langle q, e\rangle.P \mid s : h \longrightarrow P \mid s : h \cdot (p, q, v)$ $(e \downarrow v)$.

By hypothesis, $\Gamma \vdash_\Sigma s[p]!\langle q, e\rangle.P \mid s : h \rhd \Delta$. Using Lemma 3(7), (1), and (2) we have $\Sigma = \{s\}$ and

$$\Gamma \vdash s[p]!\langle q, e\rangle.P \rhd \Delta_1 \tag{8}$$
$$\Gamma \vdash_{\{s\}} s : h \rhd \Delta_2 \tag{9}$$

where $\Delta = \Delta_2 * \Delta_1$. Using 2(7) on (8) we have

$$\Delta_1 = \Delta_1', s[p] : !\langle q, S\rangle.T$$
$$\Gamma \vdash e : S \tag{10}$$
$$\Gamma \vdash P \rhd \Delta_1', s[p] : T. \tag{11}$$

From (10) by subject reduction on expressions we have

$$\Gamma \vdash v : S. \tag{12}$$

Using (QSEND) on (9) and (12) we derive

$$\Gamma \vdash_{\{s\}} s : h \cdot (p, q, v) \rhd \Delta_2; \{s[p] : !\langle q, S\rangle\}. \tag{13}$$

Using (GINIT) on (11) we derive

$$\Gamma \vdash_\emptyset P \rhd \Delta_1', s[p] : T \tag{14}$$

and then using (GPAR) on (14), (13) we conclude

$$\Gamma \vdash_{\{s\}} P \mid s : h \cdot (p, q, v) \rhd (\Delta_2; \{s[p] : !\langle q, S\rangle\}) * (\Delta_1', s[p] : T).$$

Note that $\Delta_2 * (\Delta_1', s[p] : !\langle q, S\rangle.T) \Rightarrow (\Delta_2; \{s[p] : !\langle q, S\rangle\}) * (\Delta_1', s[p] : T)$ and the consistency of $(\Delta_2 * (\Delta_1', s[p] : !\langle q, S\rangle.T)) * \Delta_0$ implies the consistency of $((\Delta_2; \{s[p] : !\langle q, S\rangle\}) * (\Delta_1', s[p] : T)) * \Delta_0$.

– [Rcv] $s[p]?(q, x).P \mid s : (q, \{p\}, v) \cdot h \longrightarrow P\{v/x\} \mid s : h$.

By hypothesis, $\Gamma \vdash_\Sigma s[p]?(q, x).P \mid s : (q, \{p\}, v) \cdot h \rhd \Delta$. By Lemma 3(7), (1), and (2) we have $\Sigma = \{s\}$ and

$$\Gamma \vdash s[p]?(q, x).P \rhd \Delta_1 \tag{15}$$
$$\Gamma \vdash_{\{s\}} s : (q, \{p\}, v) \cdot h \rhd \Delta_2 \tag{16}$$

where $\Delta = \Delta_2 * \Delta_1$. Using Lemma 2(8) on (15) we have

$$\Delta_1 = \Delta_1', s[p] : ?(q, S).T$$
$$\Gamma, x : S \vdash P \rhd \Delta_1', s[p] : T \tag{17}$$

Using Lemma 4(1) on (16) we have

$$\Delta_2 = \{s[q] : !\langle\{p\}, S'\rangle\} * \Delta_2'$$
$$\Gamma \vdash_{\{s\}} s : h \rhd \Delta_2' \tag{18}$$
$$\Gamma \vdash v : S'. \tag{19}$$

The consistency of $\Delta * \Delta_0$ implies $S = S'$. Using Lemma 5(1) from (17) and (19) we get
$\Gamma \vdash P\{v/x\} \rhd \Delta_1', s[\mathbf{p}] : T$, which implies by rule (GINIT)

$$\Gamma \vdash_\emptyset P\{v/x\} \rhd \Delta_1', s[\mathbf{p}] : T. \tag{20}$$

Using rule (GPAR) on (20) and (18) we conclude

$$\Gamma \vdash_{\{s\}} P\{v/x\} \mid s : h \rhd \Delta_2' * (\Delta_1', s[\mathbf{p}] : T).$$

Note that $(\{s[\mathbf{q}] : !\langle\{\mathbf{p}\}, S\rangle\} * \Delta_2') * (\Delta_1', s[\mathbf{p}] : ?(\mathbf{q}, S); T) \Rightarrow \Delta_2' * (\Delta_1', s[\mathbf{p}] : T)$ and the consistency of $((\{s[\mathbf{q}] : !\langle\{\mathbf{p}\}, S\rangle\} * \Delta_2') * (\Delta_1', s[\mathbf{p}] : ?(\mathbf{q}, S); T)) * \Delta_0$ implies the consistency of $(\Delta_2' * (\Delta_1', s[\mathbf{p}] : T)) * \Delta_0$.

- [Sel] $s[\mathbf{p}] \oplus \langle \mathbf{p}, l \rangle.P \mid s : h \longrightarrow P \mid s : h \cdot (\mathbf{p}, \mathbf{q}, l)$.
  By hypothesis, $\Gamma \vdash_\Sigma s[\mathbf{p}] \oplus \langle \mathbf{q}, l \rangle.P \mid s : h \rhd \Delta$. Using Lemma 3(7), (1), and (2) we have $\Sigma = \{s\}$ and

$$\Gamma \vdash s[\mathbf{p}] \oplus \langle \mathbf{q}, l \rangle.P \rhd \Delta_1 \tag{21}$$
$$\Gamma \vdash_{\{s\}} s : h \rhd \Delta_2 \tag{22}$$

where $\Delta = \Delta_2 * \Delta_1$. Using Lemma 2(2) on (21) we have for $l = l_j$ $(j \in I)$:

$$\Delta_1 = \Delta_1', s[\mathbf{p}] : \oplus\langle \mathbf{q}, \{l_i : T_i\}_{i \in I}\rangle$$
$$\Gamma \vdash P \rhd \Delta_1', s[\mathbf{p}] : T_j. \tag{23}$$

Using rule (QSEL) on (22) we derive

$$\Gamma \vdash_{\{s\}} s : h \cdot (\mathbf{p}, \mathbf{q}, l) \rhd \Delta_2; \{s[\mathbf{p}] : \oplus\langle \mathbf{q}, l\rangle\}. \tag{24}$$

Using (GPAR) on (23) and (24) we conclude

$$\Gamma \vdash_{\{s\}} P \mid s : h \cdot (\mathbf{p}, \mathbf{q}, l) \rhd (\Delta_2; \{s[\mathbf{p}] : \oplus\langle \mathbf{q}, l\rangle\}) * (\Delta_1', s[\mathbf{p}] : T_j).$$

Note that $\Delta_2 * (\Delta_1', s[\mathbf{p}] : \oplus\langle\mathbf{q}, \{l_i : T_i\}_{i \in I}\rangle) \Rightarrow (\Delta_2; \{s[\mathbf{p}] : \oplus\langle\mathbf{q}, l\rangle\}) * (\Delta_1', s[\mathbf{p}] : T_j)$ and the consistency of $(\Delta_2 * (\Delta_1', s[\mathbf{p}] : \oplus\langle\mathbf{q}, \{l_i : T_i\}_{i \in I}\rangle)) * \Delta_0$ implies the consistency of $((\Delta_2; \{s[\mathbf{p}] : \oplus\langle\mathbf{q}, l\rangle\}) * (\Delta_1', s[\mathbf{p}] : T_j)) * \Delta_0$.

- [Branch] $s[\mathbf{p}] \& (\mathbf{q}, \{L_i : P_i\}_{i \in I}) \mid s : (\mathbf{q}, \{\mathbf{p}\}, l_j) \cdot h \longrightarrow P_j \mid s : h$.
  By hypothesis, $\Gamma \vdash_\Sigma s[\mathbf{p}] \& (\mathbf{q}, \{L_i : P_i\}_{i \in I}) \mid s : (\mathbf{q}, \{\mathbf{p}\}, l_j) \cdot h \rhd \Delta$. Using Lemma 3(7), (1), and (2) we have $\Sigma = \{s\}$ and

$$\Gamma \vdash s[\mathbf{p}] \& (\mathbf{q}, \{L_i : P_i\}_{i \in I}) \rhd \Delta_1 \tag{25}$$
$$\Gamma \vdash_{\{s\}} s : (\mathbf{q}, \{\mathbf{p}\}, l_j) \cdot h \rhd \Delta_2 \tag{26}$$

where $\Delta = \Delta_2 * \Delta_1$. Using Lemma 2(12) on (25) we have

$$\Delta_1 = \Delta_1', s[\mathbf{p}] : \&(\mathbf{q}, \{l_i : T_i\}_{i \in I})$$
$$\Gamma \vdash P_i \rhd \Delta_1', s[\mathbf{p}] : T_i \quad \forall i \in I. \tag{27}$$

Using Lemma 4(3) on (26) we have

$$\Delta_2 = \{s[\mathsf{q}] : \oplus \langle \mathsf{p}, l_j \rangle\} * \Delta_2'$$
$$\Gamma \vdash_{\{s\}} s : h \triangleright \Delta_2'. \tag{28}$$

Using (GPAR) on (27) and (28) we conclude

$$\Gamma \vdash_{\{s\}} P_j \mid s : h \triangleright \Delta_2' * (\Delta_1', s[\mathsf{p}] : T_j).$$

Note that

$$(\{s[\mathsf{q}] : \oplus \langle \mathsf{p}, l_j \rangle\} * \Delta_2') * (\Delta_1', s[\mathsf{p}] : \&(\mathsf{q}, \{l_i : T_i\}_{i \in I})) \;\Rightarrow\; \Delta_2' * (\Delta_1', s[\mathsf{p}] : T_j).$$

and the consistency of $((\{s[\mathsf{q}] : \oplus \langle \mathsf{p}, l_j \rangle\} * \Delta_2') * (\Delta_1', s[\mathsf{p}] : \&(\mathsf{q}, \{l_i : T_i\}_{i \in I}))) * \Delta_0$
implies the consistency of $(\Delta_2' * (\Delta_1', s[\mathsf{p}] : T_j)) * \Delta_0$ for $j \in I$ □

Theorem 1 (Subject Reduction). *If* $\Gamma \vdash_{\Sigma} P \triangleright \Delta$ *with* Δ *consistent and* $P \longrightarrow^* P'$, *then* $\Gamma \vdash_{\Sigma} P' \triangleright \Delta'$ *for some consistent* Δ' *such that* $\Delta \Rightarrow^* \Delta'$.

Proof. Let $P \equiv \mathcal{E}[P_0]$ and $P' \equiv \mathcal{E}[P_0']$, where $P_0 \longrightarrow P_0'$ by one of the rules considered in Lemma 1. By structural equivalence we can assume $\mathcal{E} = (\vec{\nu a})(\overrightarrow{\mathsf{def}\ D\ \mathsf{in}}$ $(\vec{\nu s})([\ \ ] \mid P_1))$ without loss of generality. Theorem 2 and Lemma 3(9), (10) and (8) applied to $\Gamma \vdash_{\Sigma} P \triangleright \Delta$ give $\overrightarrow{\Gamma, a : \mathsf{G}}, \overrightarrow{X : S\,\mu\mathsf{t}.\vec{T}} \vdash_{\Sigma_0} P_0 \triangleright \Delta_0$, and $\overrightarrow{\Gamma, a : \mathsf{G}}, \overrightarrow{X : S\,\mu\mathsf{t}.\vec{T}} \vdash_{\Sigma_1} P_1 \triangleright \Delta_1$ and $\overrightarrow{\Gamma, a : \mathsf{G}}, X : S\,\mathsf{t} \vdash Q \triangleright \{y : T\}$, where $\vec{D} = \overrightarrow{X(x,y) = Q}$, $\Sigma = (\Sigma_0 \cup \Sigma_1) \setminus \vec{s}$ and $\Delta = (\Delta_0 * \Delta_1) \setminus \vec{s}$. The consistency of Δ implies the consistency of $\Delta_0 * \Delta_1$ by Lemma 3(8). By Lemma 1 there is Δ_0' such that $\overrightarrow{\Gamma, a : \mathsf{G}}, \overrightarrow{X : S\,\mu\mathsf{t}.\vec{T}} \vdash_{\Sigma_0} P_0' \triangleright \Delta_0'$ and $\Delta_0 \Rightarrow^* \Delta_0'$ and $\Delta_0' * \Delta_1$ is consistent. We derive $\Gamma \vdash_{\Sigma} P' \triangleright \Delta'$, where $\Delta' = (\Delta_0 * \Delta_1') \setminus \vec{s}$ by applying typing rules (GPAR), (GSRES), (GDEF) and (GNRES). Observe that $\Delta \Rightarrow^* \Delta'$ and Δ' is consistent.   □

References

1. Apims (2014). http://thelas.dk/index.php?title=Apims
2. Bartoletti, M., Castellani, I., Deniélou, P.-M., Dezani-Ciancaglini, M., Ghilezan, S., Pantovic, J., Pérez, J.A., Thiemann, P., Toninho, B., Vieira, H.T.: Combining behavioural types with security analysis (2014, Submitted for Journal Publication)
3. Bettini, L., Coppo, M., D'Antoni, L., De Luca, M., Dezani-Ciancaglini, M., Yoshida, N.: Global progress in dynamically interleaved multiparty sessions. In: van Breugel, F., Chechik, M. (eds.) CONCUR 2008. LNCS, vol. 5201, pp. 418–433. Springer, Heidelberg (2008)
4. Bhargavan, K., Corin, R., Deniélou, P.-M., Fournet, C., Leifer, J.J.: Cryptographic protocol synthesis and verification for multiparty sessions. In: Mitchell, J.C. (ed.) CSF 2009, pp. 124–140. IEEE Computer Society Press, Los Alamitos (2009)
5. Bocchi, L., Chen, T.-C., Demangeon, R., Honda, K., Yoshida, N.: Monitoring networks through multiparty session types. In: Beyer, D., Boreale, M. (eds.) FORTE 2013 and FMOODS 2013. LNCS, vol. 7892, pp. 50–65. Springer, Heidelberg (2013)

6. Bocchi, L., Demangeon, R., Yoshida, N.: A multiparty multi-session logic. In: Palamidessi, C., Ryan, M.D. (eds.) TGC 2012. LNCS, vol. 8191, pp. 97–111. Springer, Heidelberg (2013)

7. Bocchi, L., Honda, K., Tuosto, E., Yoshida, N.: A theory of design-by-contract for distributed multiparty interactions. In: Gastin, P., Laroussinie, F. (eds.) CONCUR 2010. LNCS, vol. 6269, pp. 162–176. Springer, Heidelberg (2010)

8. Bocchi, L., Melgratti, H., Tuosto, E.: Resolving non-determinism in choreographies. In: Shao, Z. (ed.) ESOP 2014 (ETAPS). LNCS, vol. 8410, pp. 493–512. Springer, Heidelberg (2014)

9. Bocchi, L., Yang, W., Yoshida, N.: Timed multiparty session types. In: Baldan, P., Gorla, D. (eds.) CONCUR 2014. LNCS, vol. 8704, pp. 419–434. Springer, Heidelberg (2014)

10. Bonelli, E., Compagnoni, A.B.: Multipoint session types for a distributed calculus. In: Barthe, G., Fournet, C. (eds.) TGC 2007. LNCS, vol. 4912, pp. 240–256. Springer, Heidelberg (2008)

11. Brand, D., Zafiropulo, P.: On communicating finite-state machines. J. ACM **30**, 323–342 (1983)

12. Capecchi, S., Castellani, I., Dezani-Ciancaglini, M.: Typing access control and secure information flow in sessions. Inf. Comput. **238**, 68–105 (2014)

13. Capecchi, S., Castellani, I., Dezani-Ciancaglini, M.: Information flow safety in multiparty sessions. Math. Struct. Comput. Sci. 1–43 (2015). http://journals. cambridge.org/article_S0960129514000619

14. Capecchi, S., Castellani, I., Dezani-Ciancaglini, M., Rezk, T.: Session types for access and information flow control. In: Gastin, P., Laroussinie, F. (eds.) CONCUR 2010. LNCS, vol. 6269, pp. 237–252. Springer, Heidelberg (2010)

15. Capecchi, S., Giachino, E., Yoshida, N.: Global escape in multiparty sessions. Math. Struct. Comput. Sci. 1–50 (2015). http://journals.cambridge.org/article_S0960129514000164

16. Carbone, M., Honda, K., Yoshida, N.: Structured communication-centered programming for web services. ACM Trans. Program. Lang. Syst. **34**(2), 8 (2012)

17. Carbone, M., Montesi, F.: Deadlock-freedom-by-design: multiparty asynchronous global programming. In: Giacobazzi, R., Cousot, R. (eds.) POPL 2013, pp. 263–274. ACM, New York (2013)

18. Castagna, G., Dezani-Ciancaglini, M., Padovani, L.: On global types and multiparty session. Logical Methods Comput. Sci. **8**(1), 24 (2012)

19. Castellani, I., Dezani-Ciancaglini, M., Pérez, J.A.: Self-adaptation and secure information flow in multiparty structured communications: a unified perspective. In: Carbone, M. (ed.) BEAT 2014. EPTCS, vol. 162, pp. 9–18 (2014)

20. Chen, T.-C., Bocchi, L., Deniélou, P.-M., Honda, K., Yoshida, N.: Asynchronous distributed monitoring for multiparty session enforcement. In: Bruni, R., Sassone, V. (eds.) TGC 2011. LNCS, vol. 7173, pp. 25–45. Springer, Heidelberg (2012)

21. Chen, T.-C., Honda, K.: Specifying stateful asynchronous properties for distributed programs. In: Koutny, M., Ulidowski, I. (eds.) CONCUR 2012. LNCS, vol. 7454, pp. 209–224. Springer, Heidelberg (2012)

22. Coppo, M., Dezani-Ciancaglini, M., Padovani, L., Yoshida, N.: Inference of global progress properties for dynamically interleaved multiparty sessions. In: De Nicola, R., Julien, C. (eds.) COORDINATION 2013. LNCS, vol. 7890, pp. 45–59. Springer, Heidelberg (2013)

23. Coppo, M., Dezani-Ciancaglini, M., Venneri, B.: Self-adaptive multiparty sessions. SOCA 1–20 (2014). http://dx.doi.org/10.1007/s11761-014-0171-9

24. Coppo, M., Dezani-Ciancaglini, M., Yoshida, N.: Asynchronous session types and progress for object oriented languages. In: Bonsangue, M.M., Johnsen, E.B. (eds.) FMOODS 2007. LNCS, vol. 4468, pp. 1–31. Springer, Heidelberg (2007)
25. Coppo, M., Dezani-Ciancaglini, M., Yoshida, N., Padovani, L.: Global progress for dynamically interleaved multiparty sessions. Math. Struct. Comput. Sci. 1–65 (2015). http://journals.cambridge.org/article_S0960129514000188
26. Dalla Preda, M., Giallorenzo, S., Lanese, I., Mauro, J., Gabbrielli, M.: AIOCJ: a choreographic framework for safe adaptive distributed applications. In: Combemale, B., Pearce, D.J., Barais, O., Vinju, J.J. (eds.) SLE 2014. LNCS, vol. 8706, pp. 161–170. Springer, Heidelberg (2014)
27. Dardha, O., Giachino, E., Sangiorgi, D.: Session types revisited. In: De Schreye, D., Janssens, G., King, A. (eds.) PPDP 2012, pp. 139–150. ACM Press, New york (2012)
28. Demangeon, R., Honda, K.: Nested protocols in session types. In: Koutny, M., Ulidowski, I. (eds.) CONCUR 2012. LNCS, vol. 7454, pp. 272–286. Springer, Heidelberg (2012)
29. Demangeon, R., Honda, K., Hu, R., Neykova, R., Yoshida, N.: Practical interruptible conversations: distributed dynamic verification with multiparty session types and Python. Formal Methods Syst. Des. 1–29 (2014). http://dx.doi.org/10.1007/s10703-014-0218-8
30. Deniélou, P.-M., Yoshida, N.: Buffered communication analysis in distributed multiparty sessions. In: Gastin, P., Laroussinie, F. (eds.) CONCUR 2010. LNCS, vol. 6269, pp. 343–357. Springer, Heidelberg (2010)
31. Deniélou, P.-M., Yoshida, N.: Dynamic multirole session types. In: Ball, T., Sagiv, M. (eds.) POPL 2011, pp. 435–446. ACM Press, New York (2011)
32. Deniélou, P.-M., Yoshida, N.: Multiparty session types meet communicating automata. In: Seidl, H. (ed.) Programming Languages and Systems. LNCS, vol. 7211, pp. 194–213. Springer, Heidelberg (2012)
33. Deniélou, P.-M., Yoshida, N.: Multiparty compatibility in communicating automata: characterisation and synthesis of global session types. In: Fomin, F.V., Freivalds, R., Kwiatkowska, M., Peleg, D. (eds.) ICALP 2013, Part II. LNCS, vol. 7966, pp. 174–186. Springer, Heidelberg (2013)
34. Deniélou, P.-M., Yoshida, N., Bejleri, A., Hu, R.: Parameterised multiparty session types. Logical Methods Comput. Sci. 8(4), 1–46 (2012)
35. Dezani-Ciancaglini, M., de'Liguoro, U.: Sessions and session types: an overview. In: Laneve, C., Su, J. (eds.) WS-FM 2009. LNCS, vol. 6194, pp. 1–28. Springer, Heidelberg (2010)
36. Dezani-Ciancaglini, M., Mostrous, D., Yoshida, N., Drossopoulou, S.: Session types for object-oriented languages. In: Thomas, D. (ed.) ECOOP 2006. LNCS, vol. 4067, pp. 328–352. Springer, Heidelberg (2006)
37. Fossati, L., Hu, R., Yoshida, N.: Multiparty session nets. In: Maffei, M., Tuosto, E. (eds.) TGC 2014. LNCS, vol. 8902, pp. 112–127. Springer, Heidelberg (2014)
38. Henriksen, A.S., Nielsen, L., Hildebrandt, T.T., Yoshida, N., Henglein, F.: Trustworthy pervasive healthcare services via multiparty session types. In: Weber, J., Perseil, I. (eds.) FHIES 2012. LNCS, vol. 7789, pp. 124–141. Springer, Heidelberg (2013)
39. Honda, K.: Types for dyadic interaction. In: Best, E. (ed.) CONCUR 1993. LNCS, vol. 715, pp. 509–523. Springer, Heidelberg (1993)
40. Honda, K., Hu, R., Neykova, R., Chen, T.-C., Demangeon, R., Deniélou, P.-M., Yoshida, N.: Structuring Communication with session types. In: Agha, G.,

Igarashi, A., Kobayashi, N., Masuhara, H., Matsuoka, S., Shibayama, E., Taura, K. (eds.) Concurrent Objects and Beyond. LNCS, vol. 8665, pp. 105–127. Springer, Heidelberg (2014)

41. Honda, K., Mukhamedov, A., Brown, G., Chen, T.-C., Yoshida, N.: Scribbling interactions with a formal foundation. In: Natarajan, R., Ojo, A. (eds.) ICDCIT 2011. LNCS, vol. 6536, pp. 55–75. Springer, Heidelberg (2011)

42. Honda, K., Vasconcelos, V.T., Kubo, M.: Language primitives and type disciplines for structured communication-based programming. In: Hankin, C. (ed.) ESOP 1998. LNCS, vol. 1381, pp. 122–138. Springer, Heidelberg (1998)

43. Honda, K., Yoshida, N., Carbone, M.: Multiparty asynchronous session types. In: Necula, G.C., Wadler, P. (eds.) POPL'08, pp. 273–284. ACM Press, New York (2008)

44. Hu, R., Neykova, R., Yoshida, N., Demangeon, R., Honda, K.: Practical interruptible conversations. In: Legay, A., Bensalem, S. (eds.) RV 2013. LNCS, vol. 8174, pp. 130–148. Springer, Heidelberg (2013)

45. Hüttel, H., Lanese, I., Vasconcelos, V.T., Caires, L., Carbone, M., Deniélou, P.-M., Mostrous, D., Padovani, L., Ravara, A., Tuosto, E., Vieira, H.T., Zavattaro, G.: Foundations of Behavioural Types (2014). Submitted for journal publication

46. Kouzapas, D., Yoshida, N.: Globally governed session semantics. Logical Methods Comput. Sci. **10**, 1–45 (2015)

47. Kouzapas, D., Yoshida, N., Raymond, H., Honda, K.: On asynchronous eventful session semantics. Math. Struct. Comput. Sci. **29**, 1–62 (2015)

48. Lange, J., Tuosto, E.: Synthesising choreographies from local session types. In: Koutny, M., Ulidowski, I. (eds.) CONCUR 2012. LNCS, vol. 7454, pp. 225–239. Springer, Heidelberg (2012)

49. Lange, J., Tuosto, E., Yoshida, N.: From communicating machines to graphical choreographies. In: Rajamani, S.K., Walker, D. (eds.) POPL 2015, pp. 221–232. ACM Press, New York (2015)

50. Milner, R.: Communicating and Mobile Systems: The π-Calculus. Cambridge University Press, Cambridge (1999)

51. Montesi, F., Yoshida, N.: Compositional choreographies. In: D'Argenio, P.R., Melgratti, H. (eds.) CONCUR 2013 – Concurrency Theory. LNCS, vol. 8052, pp. 425–439. Springer, Heidelberg (2013)

52. Mostrous, D., Yoshida, N., Honda, K.: Global principal typing in partially commutative asynchronous sessions. In: Castagna, G. (ed.) ESOP 2009. LNCS, vol. 5502, pp. 316–332. Springer, Heidelberg (2009)

53. Neykova, R., Bocchi, L., Yoshida, N.: Timed runtime monitoring for multiparty conversations. In: Carbone, M. (eds.) BEAT 2014. EPTCS, vol. 162, pp. 19–26 (2014)

54. Neykova, R., Yoshida, N.: Multiparty session actors. In: Kühn, E., Pugliese, R. (eds.) COORDINATION 2014. LNCS, vol. 8459, pp. 131–146. Springer, Heidelberg (2014)

55. Neykova, R., Yoshida, N., Hu, R.: SPY: local verification of global protocols. In: Legay, A., Bensalem, S. (eds.) RV 2013. LNCS, vol. 8174, pp. 358–363. Springer, Heidelberg (2013)

56. Ng, N., de Figueiredo Coutinho, J.G., Yoshida, N.: Protocols by default: safe MPI code generation based on session types. In: Franke, B. (ed.) CC 2015. LNCS, vol. 9031, pp. 212–232. Springer, Heidelberg (2015)

57. Ng, N., Yoshida, N.: Pabble: parameterised Scribble. SOCA 1–16 (2014). http://dx.doi.org/10.1007/s11761-014-0172-8

58. Ng, N., Yoshida, N., Honda, K.: Multiparty session C: safe parallel programming with message optimisation. In: Furia, C.A., Nanz, S. (eds.) TOOLS 2012. LNCS, vol. 7304, pp. 202–218. Springer, Heidelberg (2012)

59. Ng, N., Yoshida, N., Luk, W.: Scalable session programming for heterogeneous high-performance systems. In: Counsell, S., Núñez, M. (eds.) SEFM 2013. LNCS, vol. 8368, pp. 82–98. Springer, Heidelberg (2014)

60. Ng, N., Yoshida, N., Niu, X.Y., Tsoi, K.H., Luk, W.: Session types: towards safe and fast reconfigurable programming. SIGARCH CAN **40**, 22–27 (2012)

61. Nielsen, L., Yoshida, N., Honda, K.: Multiparty symmetric sum types. In: Fröschle, S.B., Valencia, F.D. (eds.) EXPRESS 2010. EPTCS, vol. 41, pp. 121–135 (2010)

62. Ocean Observatories Initiative (2010). http://www.oceanleadership.org/programs-and-partnerships/ocean-observing/ooi/

63. Padovani, L.: Deadlock and Lock Freedom in the Linear π-Calculus. In: Henzinger, T.A., Miller, D. (eds.) CSL-LICS 2014, pp. 72:1–72:10. ACM Press, New York (2014). Extended technical report available at http://hal.archives-ouvertes.fr/hal-00932356v2/document

64. Padovani, L.: Fair subtyping for multi-party session types. Math. Struct. Comput. Sci. 1–41 (2015). http://journals.cambridge.org/article_S096012951400022X

65. Pierce, B., Sangiorgi, D.: Typing and subtyping for mobile processes. J. Math. Struct. Comput. SCi. **6**(5), 409–454 (1996)

66. Benjamin, C.: Types and Programming Languages. MIT Press, Cambridge (2002)

67. Planul, J., Corin, R., Fournet, C.: Secure enforcement for global process specifications. In: Bravetti, M., Zavattaro, G. (eds.) CONCUR 2009. LNCS, vol. 5710, pp. 511–526. Springer, Heidelberg (2009)

68. Savara. SAVARA JBoss RedHat Project (2010). http://www.jboss.org/savara

69. Scribble. Scribble JBoss RedHat Project (2008). http://www.jboss.org/scribble

70. Sivaramakrishnan, K.C., Nagaraj, K., Ziarek, L., Eugster, P.: Efficient session type guided distributed interaction. In: Clarke, D., Agha, G. (eds.) COORDINATION 2010. LNCS, vol. 6116, pp. 152–167. Springer, Heidelberg (2010)

71. Swamy, N., Chen, J., Fournet, C., Strub, P.-Y., Bhargavan, K., Yang, J.: Secure distributed programming with value-dependent types. In: Chakravarty, M.M.T., Hu, Z., Danvy, O. (eds.) ICFP 2011, pp. 266–278. ACM Press, New York (2011)

72. UNIFI. International Organization for Standardization ISO 20022 UNIversal Financial Industry message scheme (2002). http://www.iso20022.org

73. Web Services Choreography Working Group. Web Services Choreography Description Language (2002). http://www.w3.org/2002/ws/chor/

74. Yoshida, N.: Graph types for monadic mobile processes. In: Chandru, V., Vinay, V. (eds.) FSTTCS 1996. LNCS, vol. 1180, pp. 371–386. Springer, Heidelberg (1996)

75. Yoshida, N., Hu, R., Neykova, R., Ng, N.: The Scribble protocol language. In: Abadi, M., Lluch Lafuente, A. (eds.) TGC 2013. LNCS, vol. 8358, pp. 22–41. Springer, Heidelberg (2014)

Refined Ownership:
Fine-Grained Controlled Internal Sharing

Elias Castegren, Johan Östlund[(✉)], and Tobias Wrigstad

Uppsala University, Uppsala, Sweden
{elias.castegren,johan.ostlund,tobias.wrigstad}@it.uu.se

Abstract. Ownership type systems give a strong notion of separation between aggregates. Objects belonging to different owners cannot be aliased, and thus a mutating operation internal to one object is guaranteed to be invisible to another. This naturally facilitates reasoning about correctness on a local scale, but also proves beneficial for coarse-grained parallelism as noninterference between statements touching different objects is easily established. For fine-grained parallelism, ownership types fall short as owner-based disjointness only allows separation of the innards of different aggregates, which is very coarse-grained. Concretely: ownership types can reason about the disjointness of two different data structures, but cannot reason about the internal structure or disjointness *within* the data structure, without resorting to static and overly constraining measures. For similar reasons, ownership fails to determine internal disjointness of *external* pointers to objects that share a common owner.

In this paper, we introduce the novel notion of refined ownership which overcomes these limitations by allowing precise local reasoning about a group of objects even though they belong to the same external owner. Using refined ownership, we can statically check determinism of parallel operations on tree-shaped substructures of a data structure, including operations on values external to the structure, without imposing any non-local alias restrictions.

1   Introduction

Ownership types [10,15] and related notions [29] have been used to solve practical problems in recent years, including simplifying concurrent and parallel programming in Java-like languages (*e.g.,* [4,12,17,27]). With ownership types, the heap is partitioned hierarchically into nested regions of memory, called *contexts* in ownership parlance. A region's contents is isolated from all external regions, giving rise to a strong notion of disjointness useful *e.g.,* to guarantee non-interference and therefore race-freedom and determinism.

Expressing disjointness between elements of *different data structures* is easy with ownership types as each object implicitly introduces a new region of memory for its representation and objects in this region cannot be pointed to from the outside. As a concrete example: if two links belong to the representation of two different lists, they cannot alias.

M. Bernardo and E.B. Johnsen (Eds.): SFM 2015, LNCS 9104, pp. 179–210, 2015.
DOI: 10.1007/978-3-319-18941-3_5

This paper addresses the inability of ownership types to express disjointness internally of a single data structure: the only way to establish disjointness is through the introduction of additional regions nested inside existing ones. For example, to express that all links in a linked list are different using ownership, each next-link must be nested inside the previous, analogous with "cons cells" in functional programming. Just like with cons cells, the nesting makes it impossible to reorder the links without also destroying the list. Likewise, encoding that each link in a linked list holds a pointer to a different element requires moving each element into the list. This has the unfortunate side-effect of forbidding any alias to them external to the list, despite the fact that the property we seek to express is of local concern only.

The underlying problem is the same in the last two cases: ownership regions lack internal structure, and the only way to add such structure is unnecessarily restricting.

In this paper we introduce the notion of *refined ownership*, which allows decomposition of a region into smaller parts. Refined ownership enables local reasoning about interference-freedom of operations on objects in different parts of a single region. Types in ownership systems are parameterised by names of external regions—so-called owner parameters. These give permission to access any object residing in that region. The idea underlying this work is that an owner parameter instead can be thought of as a set of reference permissions for *individual objects* in that memory region. Refinement reifies this set concept and allows precise local reasoning about aliasing internal to a region. Refinement allows dynamically partitioning a memory region into disjoint subsets, and propagating individual subsets to different parts of a structure. This allows expressing for example that there is only a single pointer to some external element of a structure, or that all links of a linked list point to different external objects, or that the links themselves are acyclic. Since refinement is all about local concerns, internal structure does not impose restrictions on external aliasing of the same objects.

Contributions

This paper makes the following contributions:

- We extend ownership types by means that allow capturing the internal structure of a data structure, specifically capturing how (possibly external) data is referenced from inside a data structure without imposing constraints on external references (Sect. 3).
- We demonstrate the usefulness of refined ownership on several examples of parallel applications that can be checked statically to be deterministic, including mutating maps and sorting of the elements in a list. Our examples are not possible to encode with existing ownership systems, *e.g.*, [4,11,15,18]. (Sect. 4).
- We show how refined ownership allows expressing that a chain of links is acyclic, or that a certain traversal moves within a tree-shaped subset of an otherwise aliased structure. (Sect. 4.2).

- We present a formalisation of refined ownership for a Java-like core language (Sect. 5) and formulate the key theorems (Sect. 5.8).
- We have implemented refined ownership in our prototype compiler which is briefly covered in Sect. 6.

To the best of our understanding, most existing proposals that employ ownership for structured parallelism can integrate the ideas presented here straightforwardly.

Outline. The paper proceeds as follows: Sect. 2 presents a background and partial motivation for this work. Section 3 introduces refined ownership. Section 4 additionally extends refined ownership with additional constraints for parallel programming. Section 5 presents a formal description of a core language with refinement. Section 6 briefly discusses the implementation. Section 7 discusses related work and Sect. 8 concludes.

2   Background and Motivation

In this work, we propose novel extensions to ownership types that allow dynamically partitioning a memory region (often called "an owner") into multiple disjoint subregions, and reason about the disjointness of such regions. In this section, we give a minimal introduction to object ownership to the reader, along with our motivation. A recent survey of object ownership can be found in [13].

2.1   Object Ownership

Ownership types due to Clarke *et al.* [15] partition the heap into nested regions of memory. Each object resides in a specific region and owns a private region of memory that holds its *representation objects, i.e.,* those that make up the object, such as the links of a linked list. An object may be given explicit permission to reference objects in external regions. This is captured in ownership types by parameterising types by names of external regions. The name of a region is its *owner, i.e.,* a symbolic name for the object whose representation is the region in question. Parameters of this kind are called *owner parameters*. References from outside a region to an object inside it is only allowed from the owner of that region, *i.e.,* the nearest enclosing object. The nesting structure forms a tree rooted in the outermost region World, which holds all globally accessible objects. The leftmost figure of Fig. 1 shows the nesting structure of a linked list.

2.2   Ownership-Based Effect Abstraction

Ownership is easily combined with an effect system that abstracts method behaviour as reads and writes to memory locations to allow reasoning about interference of statements in a modular fashion. Clarke and Drossopoulou [14]

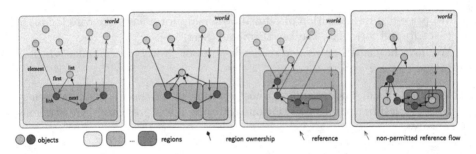

Fig. 1. First: Nesting structure of a linked list using ownership types. Second: Using static regions [1,21,32,34]. Third: Nesting to capture alias-freedom between all links in a linked list. Fourth: Extending the nesting approach to include data elements (cf. Fig. 2)

introduced the first ownership-based effect system in which methods are anno-tated by the memory regions they read and write. Ownership-based effects intro-duce a very natural notion of effect abstraction—a write to a specific object in a region R is subsumed by a "write to R", which can in turn be subsumed by a write to any region R' that encloses R. This is key to reasoning about effects deep inside a structure further up an aggregation chain where R is unknown. Several systems have adopted similar techniques for reasoning about non-interference, including Deterministic Parallel Java [4].

2.3   Ownership, Internal Structure and Disjointness

The reference structure that ownership can capture is relatively coarse-grained, and the only way to support reasoning about the innards of an object's rep-resentation is by partitioning it further in ways that severely restrict aliasing. Ownership shines when expressing the separation of different objects' represen-tations, which works well for proving non-interference of operations on *different data structures*. However, when manipulated objects belong to a single region, ownership offers little reasoning power to distinguish between them. With respect to ownership, two pointers into a single memory region are always potentially aliases. This limits the usefulness of ownership for performing parallel operations on a data structure, or parallely permuting the internals of a data structure.

We now survey several ways offered in the ownership literature to overcome this problem. These solutions are all unsatisfactory; they either require the num-ber of disjoint regions to be known statically, or introduce severe restrictions on aliasing, both internal and external, and make structures difficult to change. Figure 1 shows a graphic overview of the situations below, and how additional structure through nesting introduces additional restrictions.

"Vanilla" Linked List. Figure 1(First) shows the canonical ownership types example—a single-linked list. Each link is owned by the list object, which is depicted by their nesting inside the region owned by the list. It is possible to distinguish two links of different lists, but not two links of a single list.

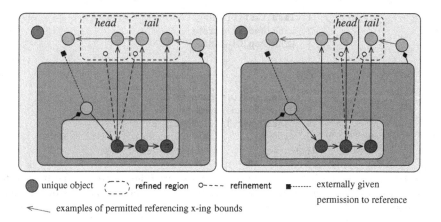

Fig. 2. Memory structure with refined regions. Different links are permitted to reference different subsets of the data domain. Note that unique objects are not part of any region.

Links in Different Regions. Figure 1(Second) shows placement of links inside different regions [1, 21, 32, 34] of a single list. In a system such as this, each object can introduce a statically known number of representation regions. With each link in a different region, the individual links are clearly distinguishable, but limited to a statically known number of regions, and thus links.

Nesting of Links. Figure 1(Third) shows the use of nesting to establish disjointness of each link. Such examples can be found in the literature, for example in DPJ [2]. This avoids a statically known number of regions, but prevents standard tricks such as last links, and makes insertion and deletion of links impossible since this affects nesting structure.

Nesting of Links and Data Elements. Figure 1(Fourth) shows a similar case to Fig. 1(Third), where also the *elements* are nested in the list. This enables establishing disjointness between links, and also between elements in the list. However, it is a very strong restriction that prevents external pointers to the elements of the list, including banning elements of being part of several structures at once.

We now present our extension of ownership, which overcomes all of these problems in an entirely static system.

3   Ownership Refinement

The idea underlying this work is that of an owner as a *set of reference permissions*. Normally, an owner P is treated as a token granting permission to reference all the objects owned by P. Ownership refinement allows the treatment of an owner P as a set of individual permissions for objects owned by P,

```
1  class List[D] {
2    Link[Rep,D] first;
3    Link[Rep,D] last;
4    ...
5  }
6
7  class Link[D = Head ⊎ Tail] {
8    Object[Head] element;
9    Link[Owner,Tail] next;
10   ...
11 }
```

Fig. 3. Parts of a linked list with ownership refinement. The permission set D given to the list object is split dynamically over the links so that each link gets its own unique permission set (cf. Fig. 2).

i.e., in P's representation. A set of permissions may be split into several disjoint subsets that can be used in different parts of a data structure for more precise reasoning about aliasing, hence the name ownership refinement. Refinement is a simple idea with important consequences. For example, if a binary tree is para- metrised by the owner P as the owner of the elements stored in the tree, we can split P into three disjoint sets, one passed down to the left subtree, one to the right subtree, and one kept in the current node to reference its element. Each new node created inside the tree becomes a new witness of the existence of an additional subregion, which is what allows dynamic partitioning (similar to the nesting of links in Fig. 1, but horizontal rather than nested). This naturally captures that no element can be found in both subtrees of a tree.

As a concrete example Fig. 3 shows part of the code for a linked list that uses ownership refinement. The clause D = Head ⊎ Tail in the class head introduces two disjoint permissions sets Head and Tail whose union is D. We call Head and Tail *refined owners*. The permission set Head is used in the element field, and Tail is passed on to the next link, being bound to its D and further refined.

With the introduction of ownership refinement arises an important distinc- tion between global and local ownership information. Traditionally, ownership types is only concerned with global properties, meaning that each piece of owner- ship information reflects some global nesting invariant in the system, *e.g.*, region P is nested inside region Q. In contrast, *refined owners are concerned with how aliases are used locally, inside some data structure*. Different objects may refine a common region in different ways, *e.g.*, a linked list and a binary tree may share elements from the same global region P, but internally partition P differently depending on the requirements for each particular case.

3.1   Adding Values to Refined Owners

So far, we have only talked about how regions can be partitioned and partitions "forwarded" to different sub-objects to structure an aggregate, but not about how regions can be populated. We currently support three ways of mapping objects to refined owners:

Unique References. The simplest one involves unique references. If a reference is unique, then it can trivially be cast (and consumed) into a refined owner since there can be no other reference to the same object from some other refinement. Once the reference has been cast into the refined owner, the object can be freely aliased externally—uniqueness is simply a way to guarantee well-formed construction of refined regions.

Internal Creation. Nothing prevents objects from being created inside a refined owner, *e.g.,* `new Object[Head]();`. Once created, the object may be shared and aliased freely externally—under the regular rules of ownership types. (This is the approach taken in [3].)

Copies from Disjoint Regions. If regions P and Q are disjoint, then we may take any element of Q and add it to P. This allows copying elements by reference across collections with preserved disjointness, but as soon as one element from Q is moved to P, P and Q are no longer disjoint.

3.2   Copying Elements Across Lists

Figure 4 shows the implementation of a copy method that creates a new list sharing its elements with the original, just as promised in the contributions. Copying is initiated by a call to `copyList`, which creates a new list object passing the current `first` link as argument. The constructor of `List` calls `copyLink` on the link argument to produce a copy of the links, sharing the same elements. Generalising this code is possible *e.g.,* by passing in a factory object to `copyList`, which knows how to create both lists and links (or equivalent). Note that we currently only support cloning from structures with equal or less refinement, *i.e.,* we can copy a tree to a list but not vice versa. We are currently working on solving this problem in a way elegant enough for actual use.

Moving things into a refined structure loses uniqueness. However, uniqueness is key to manipulating structures with refined ownership, and transforming them without breaking the disjointness invariants on the refined owners. This is explained in the upcoming section.

3.3   Manipulating Structures with Refined Ownership

One of our criticisms of the use of plain ownership types to introduce internal disjointness in Sect. 2.3 is that the nesting structure is fixed, which precludes things like last pointers and reordering of links. As refined ownership gives a unique type to each object in a tree, one might suspect that manipulation of internal structure would be just as difficult. For example, what is the type of `next.element` in a link in Fig. 3? Since we cannot externally name the refined owner of the next link, we have two choices—to ban this access altogether, or come up with a way of abstracting the ownership information so that it becomes well-typed. We choose the latter, and widen the return type of `next.element`

```
1  // In List class
2  [X] List[X,D] copyList() reads Rep, D {
3    return new List[X,D](first);
4  }
5
6  [X] List(Link[X,D] o) {
7    first = o.copyLink[This]();
8  }
9
10 // In Link class
11 [X] Link[X,D] copyLink() reads Rep, D {
12   return new Link[X](element,
13     next == null ? null : next.copyLink());
14 }
15
16 [H, T] Link(Object[H] e, Link[Owner,T] n) {
17   this.element = e;
18   this.next    = n;
19 }
```

Fig. 4. Copying elements across collections with preserved information about alias-freedom. Method in `List` to the left and `Link` to the right. `X` is implicitly nested inside `World`.

to `Object[Tail]`. Note that this `Tail` is the one used to type `next` and which is called `D` and refined into another `Head` and `Tail` inside the next `Link`. Thus the first `Tail` includes the new `Head` and this is why we can widen the type.

In the general case, we cannot assign to a field whose type uses a refined owner, as that could violate the tree-shapedness of the object graph (*e.g.*, `next.next = next`). To facilitate manipulation of internal structure, we allow a program to mediate between different views of a tree. So far, trees have been regular object structures whose members have been guaranteed to be locally unique.

Relying on refined ownership, we can introduce a novel "supercharged version" of the focus construct from the work of DeLine and Fähdrich [19] that allows us to temporarily view all references in a tree as unique references which must be treated linearly. The main difference between our focus operation and that of DeLine and Fähdrich is that focusing on a root variable makes all parts of *all trees rooted at that variable unique*, and not just the root variable itself. Thus, our focus is "deep"—it focuses on an entire nested tree-shaped substructure of the heap and forbids access to all aliases to the substructure that are not in the tree; for example, if we focus on `first` of a linked list, we are prevented from reading `last` at the same time.[1]

---

[1] We would like to note that there's risk of confusion as to which tree is focused on when focusing on an owner that is used multiple times in a type. The solution is simple and just requires that the programmer is explicit about which parameter is meant, however we leave this out in the examples and the formalism because it does not add to the story.

```
 1  class Link[D = Head ⊎ Tail] {
 2    Object[Head] element;
 3    Link[Owner, Tail] next;
 4
 5    Link(Object[D] e, Link[Owner, D] n) focuses D {
 6      this.element = consume e;
 7      this.next    = consume n;
 8    }
 9
10    Link[Owner, D] unlink(int i) focuses D {
11      return i==1 ? consume next : next.unlink(i-1);
12    } }
13
14  class List[D] {
15    Link[Rep, D] first = null;
16    int length = 0;
17
18    void prepend(unique Object[D] e) writes Rep {
19      focus D in first {
20        first = new Link(consume e, consume first);
21      }
22      this.length++;
23    }
24
25    void sort() writes Rep {
26      focus D in first {
27        first = mergesort(consume first, this.length); } }
28
29    Link[Rep, D] mergesort(Link[Rep, D] l, int length) focuses D {
30      if (length < 2) { return consume l; }
31      Link[Rep, D] second = l.unlink(length/2);
32      finish {
33        async { l = mergesort(consume l, length/2) }
34        async { second = mergesort(consume second, length/2 + length % 2) }
35      }
36      return merge(consume l, consume second);
37    }
38
39    Link[Rep, D] merge(Link[Rep, D] f, Link[Rep, D] s) focuses D {
40      if (f == NULL) {
41        return consume s;
42      } else if (s == NULL) {
43        return consume f;
44      } else if (f.element < s.element) {
45        f.next = merge(consume f.next, consume s);
46        return consume f;
47      } else {
48        s.next = merge(consume f, consume s.next);
49        return consume s;
50      } }
51    }
```

Fig. 5. A parallel merge sort using the focus mode. The novel **focuses** effect annotation captures the permission set of the tree being permuted (see Sect. 4.3).

The assignment `this.element = next.element` is not well-typed outside of a focus, as the types of `this.element` and `next.element` are not assignment compatible. During a focus, `this.element` and `next.element` are simply two unique pointers to leaves of the same tree, and are therefore assignment compatible. It is easy to see that a linear manipulation of a tree will preserve tree integrity.

Figure 5 shows a linked list with a single tree rooted in `first`, which guarantees that all elements of the list are mutually unique, and that the list is acyclic. The `sort()` method uses an implementation of mergesort. The `focus` operation in `sort()` changes the type of all links reachable from `first` into `unique Link[Rep, D]`, and the type of all elements into `unique Object[D]`, where D is the unrefined owner D in `List`. The mergesort implementation is very straightforward, and uses a fair amount of destructive reads which are explicated through an optional `consumes` annotation[2]. Apart from the method-level annotation `focuses D`, which is needed to capture the fact that we are operating inside this "linear mode", `mergesort()` and `merge()` have no effects and are therefore safe to run in parallel. Methods with focus are only callable from inside focus blocks or other focusing methods.

4   Refined Ownership for Parallelism

The need for refined ownership arose in the context of our Joelle programming language [12,31], which relies on ownership information for isolation to preserve active object integrity. In this work, we have found the need to express additional parallel possibilities internal to an active object method, especially to coordinate work on different parts of shared structures, which is not possible across active object isolates.

Refined ownership allows expressing what objects may be referenced from what internal part of an aggregate object by explicating how permissions are propagated to different subparts of a structure. As a side-effect, this propagation also exposes the internal structure of a set of objects.

For the reasons outlined in Sect. 2, existing ownership systems are not a good basis for effect systems with the goal of operating in parallel on objects belonging to a single structure. We now show how refined ownership overcomes these limitations and enables statically checking task-based parallelism in the fork-join style on elements inside a collection, similar to the proposed extension of DPJ [3], which supports collections with external data.

Deterministic parallel operation on elements of a collection in fork-join style requires that tasks running in parallel do not interfere. Similar to other systems, this amounts to the absence of read–write or write–write conflicts in the tasks' effect footprints. A special case of this is that we must make sure that no task is spawned twice.

---

[2] The formalism notably uses explicit destructive reads and unique pointers for simplicity, but these can be inferred in the actual language implementation. See however anecdotal evidence by Gordon *et al.* [20] that programmers appreciate explicit operations on uniques.

```
1    class BST[Owner, D] where Owner # D {
2      Node[Rep, D] root = null;
3
4      void insert(unique Object[D] el) writes Rep {
5        if (root == null) root = new Node(consume el);
6        else root.insert(consume el);
7      }
8
9      void pmap(Fun[World] f) writes D {
10       if (root == null) return;
11       root.pmap(f);
12     }
13   }
14
15   class Node[D = Left ⊎ Pivot ⊎ Right] where Owner # D {
16     Object[Pivot] element;
17     Node[Owner, Left] left;
18     Node[Owner, Right] right;
19
20     Node(unique Object[D] el) writes Rep {
21       this.element = consume el;
22     }
23
24     void insert(unique Object[D] el) writes Owner {
25       if (el > this.element) {
26         if (right == null) right = new Node(consume el);
27         else right.insert(consume el);
28       } else {
29         if (left == null) left = new Node(consume el);
30         else left.insert(consume el);
31       }
32     }
33
34     void pmap(Fun[World] f) reads Owner, writes D {
35       finish {
36         async { if (left != null) left.pmap(f); }
37         async { f.call(this.element); }
38         async { if (right != null) right.pmap(f); }
39       }
40     }
41   }
42
43   interface Fun {
44     [X] void call(Object[X] o) writes X;
45   }
```

Fig. 6. Parts of a a binary search tree with ownership refinement. The permission set D given to the tree object is split dynamically over the nodes so that each node gets its own unique permission set. As a side-effect, all nodes in the tree have unique types, which guarantees that the tree really is tree-shaped. The **where**-clause introduces the requirement that two owners may not be instantiated by overlapping owners, which can otherwise easily happen due to reflexivity of owner relations.

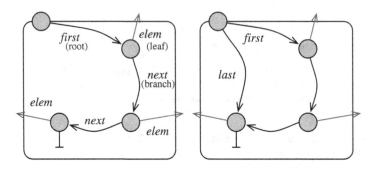

Fig. 7. Two examples of linked lists with different internal trees. Left: single-linked list with all first/next references mutually unaliased. Right: adds a last reference to the leftmost example that potentially aliases any link in the list.

4.1   A Motivating Example

Figure 6 shows how refined ownership can express that each node in a binary search tree holds a set of permissions to reference external objects which are disjoint from the permissions of its subtrees. Each tree node is parameterised over some set of permissions that is split into three—one part that is kept and used to type the element reference, and two that are passed on to the subtrees. The fact that the permission sets of the branches are disjoint and smaller than their common root establishes a tree-order and precludes cycles or dag-like structures. Together, these two properties are enough to guarantee the *deterministic execution* of a parallel operation on the list, *e.g.*, `map`, as long as the parallel operation has no write effects other than to the element.

The `insert` method in BST takes a unique argument and inserts it in the tree. The owner `unique[D]` denotes that the value is not aliased by any other variable or field in the system, and the parameter D is necessary to bound how the value may flow outwards in the nesting hierarchy[3]. In the different branches in `insert` in the `Node` class, the unique value is *moved* into a refined owner—Left or Right.

Using the disjointness information from the refined ownership, we can straightforwardly type the `pmap` method in BST that performs a parallel traversal of the tree rooted in `root` and performs an in-place update. It takes a function parameter object specified by the `Fun` interface whose `call` function is parameterised with the name of a region that gives it temporary permission to access and modify one isolated element (cf. *e.g.*, [11]) in the list, and the effect annotation `writes X` ensures that no shared data is touched.

Ownership refinement is able to statically capture that this implementation is interference-free: apart from a read effect to `Owner`, the effects of lines 43–45 are `writes Left`, `writes Pivot` and `writes Right`, which are disjoint and

---

[3] This is called the movement bound and investigated further in [11]. It is required for soundness of ownership transfer, but otherwise of little relevance for refined ownership.

subsumed by a write effect to D, which is visible externally. Thus, the three async blocks (inspired by [9]) are non-interfering and can be wrapped in the same finish[4].

Unsound attempts at using the internal parallelism of two data structures (*e.g.,* parallel uses of iterate of two BST's with possibly overlapping data elements) will be detected at the outermost call-site by the effect system and rejected.

4.2   Tree-Shaped Subgraph Traversals

The key to statically checking determinism of parallel operations on objects with refined ownership is to prove that data is accessed in the shape of a tree. Each node in the tree has an element that is unique in the tree and one or more subtrees with elements which are unique to them.

Figure 7 shows two linked lists. The brown and blue arrows are branches and the green arrows are leaves. The first branch is a root. The leftmost list has just one single "tree" rooted in first of next-links (brown arrows) pointing to a set of locally unique elements (green arrows). The right list has two trees rooted in first and last, but is otherwise the same. This shows that "weakly overlapping trees" may exist simultaneously in the same aggregate.[5]

4.3   Constructing Trees for Parallel Programming

Tree-shaped data structures and tree-shaped computation are recurring in parallel programming and structured parallelism where a divide-and-conquer approach is common.

Roots, branches and *leaves* are the building blocks of trees that are possible to operate on in parallel in a statically checkable fashion. Roots identify starting points of a tree and are the cut-off points inside which parallel operations are deterministic. Branches build tree-shaped paths to preclude interference of parallel tasks. Allowing paths that are cyclic or dags might lead to tasks racing on some object due to an unfortunate alias to a shared object. In the case of collections, leaves are often the objects of most interest, *i.e.,* the elements mutated in parallel.

Refined ownership provides a simple way to statically identify roots, branches, leaves, or combinations. The use of a refined owner in the owner position—the first position—of a type denotes a leaf, because it cannot be further refined internally. A leaf sits on no other path in the tree and may safely be operated on in

---

[4] In addition to disjoint effects, two async blocks in the same finish must not update the same variable—a trivial compile-time analysis.

[5] The tree rooted in last in Fig. 7 notably stops immediately after its first object. To create trees which overlap in "interesting ways," *e.g.,* a doubly-linked lists additional machinery is needed, *e.g.,* a "splitting operator" that allows introduction of multiple aliases to a value which are bound to different trees. We do not yet have a convincing *simple* extension for this, and refer to future work.

parallel. For example, Line 19 in Fig. 6 exemplifies this in the type of `element`. Binding a refined owner to an owner parameter that is further refined denotes a branch—more values which are unique in the same tree (may be) are available in the substructure of an object with such a type. This is exemplified by the `next` fields of links, which bind `Tail` to the parameter D of some link, which is then further refined into the (different) `Head` and `Tail` subsets of the next link. Finally, binding a non-refined owner to an owner parameter that is refined denotes a root, as exemplified by `first`. It is possible for a field to be a root in one tree and *e.g.,* a branch in another tree at the same time, although that is not in the example.

From this point forward, we will use the term *tree of A* to denote a connected subgraph of an object graph where all paths to objects in region A are alias-free. When A is not important, we will simply write *tree*.

5   A Type System for Refined Ownership

We now turn to the formal description of refined ownership in the context of a simple Java-like core language building on [11,33]. There are three main components to our formalisation:

1. ownership regions that are nested (inside $\prec^*$, outside $\succ^*$),
2. refinement which splits regions into disjoint subregions ($\uplus$), and
3. ownership-based effects and reasoning about disjointness ($\#$) of effects (see [14]).

Refinement partitions memory "horizontally" whereas nesting partitions memory "vertically". There is a correlation between *static regions* in the form of [1,21,32,34] and refinement. In a way, refinement can be seen as a generalisation of static regions by allowing the partitioning into regions dynamically, including partitioning of external regions.

The syntax of our core language is shown in Fig. 8. The type environment Γ records the usual information for an ownership system: types of local variables and names of the memory regions in scope and their nesting relations. In addition, Γ also includes refinement relations, *e.g.,* Data is refined into `Head` and `Tail` in the `Link` class in Fig. 3, as well as disjointness assertions, *e.g.,* `Head` # `Tail`. In the code examples refinement declarations are inlined in the parameter declarations of classes. For simplicity, we have separated them out in a `where`-clause in the formalism. This is a purely syntactic difference.

A program is a list of class definitions excluding the empty top-level class `Object`, which is implicitly defined. Classes contain field and method definitions, are parameterised over permissions to reference external regions and derive from a single super class. Owner parameters may be refined into multiple disjoint regions. Methods also take zero or more owner parameters—temporary permissions to reference objects—and are annotated with effect declarations. The `focuses` annotation on methods is not a computational effect. Focused

$$P ::= \overline{C} \textbf{ class } \texttt{Object} \{ \} \qquad\qquad\qquad\qquad \textit{Program}$$
$$C ::= \textbf{class } \texttt{C}[\textbf{Owner}, \overline{p}] \textbf{ where } \overline{K} \textbf{ extends } \texttt{D}[\overline{q}] \{\overline{F} \, \overline{M}\} \qquad \textit{Class Declaration}$$
$$F ::= T \, F \qquad\qquad\qquad\qquad\qquad\qquad\qquad\qquad \textit{Field}$$
$$M ::= [\overline{p}] \, T \, \texttt{m}(\overline{T \, x}) \, \overline{E} \, \{ \, s \, ; \textbf{return } a \, \} \qquad\qquad \textit{Method}$$
$$\quad | \quad [\overline{p}] \, T \, \texttt{m}(\overline{T \, x}) \textbf{ focuses } p \, \{ \, s \, ; \textbf{ return } a \, \}$$
$$k ::= \textbf{Owner } | \textbf{ Rep } | \textbf{ World } | \, p \qquad\qquad\qquad \textit{Owners}$$
$$\quad | \quad p \textbf{ from } q \, | \textbf{ focused } p \, | \, \boxed{k_{\iota}}$$
$$K ::= k \prec^* k \, | \, k \succ^* k \, | \, k \# k \, | \, p = q_1 \uplus \ldots \uplus q_n \qquad \textit{Constructor}$$
$$T ::= \texttt{C}[\overline{k}] \, | \textbf{ unique } \texttt{C}[\overline{k}] \qquad\qquad\qquad\qquad \textit{Types}$$
$$E ::= \textbf{wr } p \, | \textbf{ rd } p \qquad\qquad\qquad\qquad\qquad\quad \textit{Effect}$$
$$s ::= cmd \, | \quad cmd \, ; s \, | \textbf{ finish}\{\overline{par}\} \, ; s$$
$$\quad | \quad \textbf{focus } p \textbf{ in } x \textbf{ as } y \, \{s\} \, ; s \qquad\qquad\qquad \textit{Statements}$$
$$par ::= \textbf{async}\{s\} \qquad\qquad\qquad\qquad\qquad\qquad \textit{Parallelism}$$
$$cmd ::= \textbf{skip } | \, T \, x = e \, | \, path = e \qquad\qquad\qquad \textit{Commands}$$
$$\quad | \quad x = path \, . \, \texttt{m}[\overline{k}] \, (\overline{a}) \qquad\qquad\qquad\qquad \textit{Expressions}$$
$$e ::= \textbf{null } | \, a \, | \, (T) \, a \, | \textbf{ new } T$$
$$a ::= path \, | \textbf{ consume } path \qquad\qquad\qquad\qquad \textit{Arguments}$$
$$path ::= t \, | \, t.f \qquad\qquad\qquad\qquad\qquad\qquad\quad \textit{Paths}$$
$$t ::= x \, | \textbf{ this } | \, \boxed{v} \qquad\qquad\qquad\qquad\qquad\quad \textit{lvalues}$$

Fig. 8. Core language syntax. We follow FJ [22] and write $\overline{p}$ to mean sequences of p, etc. Gray blocks denote syntax only appearing in the running program.

methods deal exclusively with permutation of one specific tree, and are able to run in parallel as shown in Fig. 5.

The keywords World, Owner and Rep denote special owners, *i.e.,* memory regions: World denotes the globally accessible top-most owner, Owner is the compulsory name of the first owner parameter of a class, which denotes the region in which the object resides. The implicitly introduced owner Rep denotes the object's own internal region, which is nested directly inside Owner. Types use the form $A \, \texttt{C}[\overline{p}]$ where the parameters $\overline{p}$ are names of regions whose objects the object may reference. The first parameter also denotes the owning region. A is an optional uniqueness annotation for values pointed to by a single pointer in the entire system or values belonging to the currently permuted tree respectively. To keep matters simple, we employ destructive reads to preserve uniqueness, using consume.

5.1 Typing of Statements

Statement type rules have the form $P; \Gamma \vdash_p s \triangleright \overline{E} \dashv \Gamma'$ which denotes that the statement s is well-formed under Γ, has effects $\overline{E}$, and results in a new Γ', which is possibly extended by new variable declarations.

The subscripted owner name on the turnstile tracks the tree in focus (possibly none, denoted ϵ). To clarify several aspects of refinement and focusing, we split certain rules for looking up fields, etc. into several rules depending on the receiver (this or not), and whether the accessed value is part of the focused tree or not. The three different versions of look-up are exercised in S-UPDATE-FIELD, S-UPDATE-FOCUSED-FIELD, and S-UPDATE-FIELD-INTERNAL.

These rules follow similar patterns, but the differences are important: reading fields and calling methods with this as the receiver (S-UPDATE-FIELD-INTERNAL), we do not need to widen types to hide refinement (which we must do for external field look-up). Furthermore, we can allow access to types which use Rep as an owner (this is the normal ownership containment invariant). During a focus on some p, we must treat values in the tree of p linearly, which is implemented by giving them unique types in these rules. Field lookups (also external method calls) must respect ownership containment (types using this must not be in the signature), and, in the case of a read, any internal refinement must be widened, *e.g.,* from Head to Data in Fig. 3. In the case of a write, such accesses are denied as they are unsound.

Without loss of generality, we restrict the focus construct (S-FOCUS) to operate only on local variables. Focusing on a tree in p rooted in some variable x introduces a new variable y that aliases x, but with a different type, where p is replaced by the special owner focused p. This allows us to restrict the lifetime of the focus even in the presence of owner-polymorphic methods, and also gives us a simple means of identifying the currently focused tree. The helper function owners extracts the owner parameters from a type. Finally, we require that the statements inside the block produce no effects, which is the case for manipulations on a focused structure.

Calling a method (S-CALL) substitutes the owners of the types of the method with the receiver's owners at the call-site plus any owner parameters and checks that the types of arguments and result variable are correct. To avoid cluttering the rules, this is handled inside mLookup. Calls to focused methods (S-CALL-FOCUSED) additionally require that the method's focused owner corresponds to the current owner in focus.

For simplicity, we disallow calling methods and reading and writing fields on globally unique values since their main use here is to install values in refined regions. This is handled in the look-up functions, which take the target type as argument.

The full treatment of uniques requires a well-understood borrowing construction (for example [6,11,38]), but this falls out of scope for this paper where uniqueness is mainly used for storing in refined subregions. This loses uniqueness after which the value can be used in a normal fashion.

Our formalisation of async–finish style parallelism in (PAR-ASYNCS) and (PAR-ASYNC) is inspired by [23]. Async blocks inside a common finish may not have conflicting effects or destructively read or assign the same variables. The function vars extracts such variables from the statements of a par block.

$$\boxed{P; \Gamma \vdash_p s \rhd \overline{E} \dashv \Gamma'} \qquad\qquad\qquad \textit{(statements)}$$

S_UPDATE_FIELD

$$\frac{p \notin k_1, \overline{k} \qquad \Gamma(x) = \mathsf{C}[k_1, \overline{k}] \qquad \mathsf{fLookupWr}\,(P, \mathsf{C}[k_1, \overline{k}], f) = T \qquad P; \Gamma \vdash_p e : T \rhd \overline{E}}{P; \Gamma \vdash_p x.f = e \rhd \overline{E}, \mathbf{wr}\, k_1 \dashv \Gamma}$$

S_UPDATE_FIELD_INTERNAL

$$\frac{p \notin \overline{k} \qquad \Gamma(\mathrm{this}) = \mathsf{C}[\overline{k}] \qquad \mathsf{fLookupInt}\,(P, \mathsf{C}[\overline{k}], f) = T \qquad P; \Gamma \vdash_p e : T \rhd \overline{E}}{P; \Gamma \vdash_p \mathrm{this}.f = e \rhd \overline{E}, \mathbf{wr}\, \mathbf{Rep} \dashv \Gamma}$$

S_UPDATE_FOCUSED_FIELD

$$\frac{p \in \overline{k} \qquad \Gamma(x) = \mathsf{C}[\overline{k}] \qquad \mathsf{fLookupRd}\,(P, \mathsf{C}[\overline{k}], f) = T \qquad P; \Gamma \vdash_p e : T \rhd \overline{E} \qquad \mathsf{owners}\,(T) = \overline{k}}{P; \Gamma \vdash_p x.f = e \rhd \overline{E} \dashv \Gamma}$$

S_UPDATE_FOCUSED_FIELD_INTERNAL

$$\frac{\Gamma(\mathrm{this}) = \mathsf{C}[\overline{k}] \qquad \mathsf{fLookupInt}\,(P, \mathsf{C}[\overline{k}], f) = T \qquad P; \Gamma \vdash_p e : T \rhd \overline{E} \qquad p \in \overline{k} \qquad \mathsf{owners}\,(T) = \overline{k} \qquad p \in \overline{k}}{P; \Gamma \vdash_p \mathrm{this}.f = e \rhd \overline{E} \dashv \Gamma}$$

S_FOCUS

$$\frac{\Gamma(x) = T \qquad p \in \mathsf{owners}\,(T) \qquad T\{\mathbf{focused}\, p/p\} = T' \qquad \Gamma' = \Gamma, y : T' \qquad P; \Gamma' \vdash_p s_1 \rhd \epsilon \dashv \Gamma_1 \qquad P; \Gamma \vdash_\epsilon s_2 \rhd \overline{E} \dashv \Gamma''}{P; \Gamma \vdash_\epsilon \mathbf{focus}\, p \,\mathbf{in}\, x \,\mathbf{as}\, y \,\{s_1\}\,;\, s_2 \rhd \overline{E} \dashv \Gamma''}$$

S_CALL

$$\frac{P; \Gamma \vdash_{\_} x : T \rhd \epsilon \qquad P; \Gamma \vdash_{\_} y : T' \rhd \epsilon \qquad \overline{P; \Gamma \vdash_p a : T \rhd \overline{E}}^{\,i} \qquad \mathsf{mLookup}\,(P, T', m, \overline{k}) = [\overline{k}]\, T\, m(\overline{T\, x}^{\,i})\overline{E}''\{s;\, \mathbf{return}\, a\}}{P; \Gamma \vdash_p x = y.m[\overline{k}](\overline{a}^{\,i}) \rhd \overline{E}'', \overline{\overline{E}}^{\,i} \dashv \Gamma}$$

S_CALL_INTERNAL

$$\frac{P; \Gamma \vdash_{\_} x : T \rhd \epsilon \qquad P; \Gamma \vdash_{\_} \mathrm{this} : B \rhd \epsilon \qquad \overline{P; \Gamma \vdash_p a : T \rhd \overline{E}}^{\,i} \qquad \mathsf{mLookupInt}\,(P, B, m, \overline{k}) = [\overline{k}]\, T\, m(\overline{T\, x}^{\,i})\overline{E}''\{s;\, \mathbf{return}\, a\}}{P; \Gamma \vdash_p x = \mathrm{this}.m[\overline{k}](\overline{a}^{\,i}) \rhd \overline{E}'', \overline{\overline{E}}^{\,i} \dashv \Gamma}$$

S_CALL_FOCUSED

$$\frac{P;\Gamma \vdash_{\_} x : T \triangleright \epsilon \qquad P;\Gamma \vdash_{\_} y : T' \triangleright \epsilon \qquad \overline{P;\Gamma \vdash_p a : T \triangleright \overline{E}}^{\,i} \qquad \mathsf{mLookup}\,(P, T', m, \overline{k}) = [\overline{k}]\,T\,m(\overline{T\,x}^{\,i})\,\textbf{focuses}\,p\{s;\,\textbf{return}\,a\}}{P;\Gamma \vdash_p x = y.m[\overline{k}](\overline{a}^{\,i}) \triangleright \epsilon, \overline{\overline{E}}^{\,i} \dashv \Gamma}$$

S_CALL_FOCUSED_INTERNAL

$$\frac{P;\Gamma \vdash_{\_} x : T \triangleright \epsilon \qquad P;\Gamma \vdash_{\_} this : B \triangleright \epsilon \qquad \overline{P;\Gamma \vdash_p a : T \triangleright \overline{E}}^{\,n} \qquad \mathsf{mLookupInt}\,(P, B, m, \overline{k}) = [\overline{k}]\,T\,m(\overline{T\,x}^{\,n})\,\textbf{focuses}\,p\{s;\,\textbf{return}\,a\}}{P;\Gamma \vdash_p x = this.m[\overline{k}](\overline{a}^{\,n}) \triangleright \epsilon, \overline{\overline{E}}^{\,n} \dashv \Gamma}$$

S_VARIABLE_DECLARATION

$$\frac{x \notin \mathsf{dom}\,(\Gamma) \qquad P;\Gamma \vdash_p e : T \triangleright \overline{E}}{P;\Gamma \vdash_p T\,x = e \triangleright \overline{E} \dashv \Gamma, x : T}$$

S_UPDATE_LOCAL

$$\frac{\Gamma(x) = T \qquad P;\Gamma \vdash_p e : T \triangleright \overline{E}}{P;\Gamma \vdash_p x = e \triangleright \overline{E} \dashv \Gamma}$$

S_SEQUENCE

$$\frac{P;\Gamma_1 \vdash_p cmd \triangleright \overline{E}_1 \dashv \Gamma_2 \qquad P;\Gamma_2 \vdash_p s \triangleright \overline{E}_2 \dashv \Gamma_3}{P;\Gamma_1 \vdash_p cmd; s \triangleright \overline{E}_1, \overline{E}_2 \dashv \Gamma_3}$$

PAR_ASYNCS

$$\frac{P;\Gamma \vdash_{\_} par \triangleright \overline{E} \dashv \Gamma' \qquad P;\Gamma' \vdash_{\_} s \triangleright \overline{E}' \dashv \Gamma'' \qquad P;\Gamma \vdash \overline{E} \;\#\; \overline{E}' \qquad \mathsf{vars}\,(par) \cap \mathsf{vars}\,(s) = \emptyset}{P;\Gamma \vdash_{\_} par\,\textbf{async}\,\{s\} \triangleright \overline{E}, \overline{E}' \dashv \Gamma}$$

PAR_ASYNC

$$\frac{P;\Gamma \vdash_{\_} s \triangleright \overline{E} \dashv \Gamma'}{P;\Gamma \vdash_{\_} \textbf{async}\,\{s\} \triangleright \overline{E} \dashv \Gamma}$$

S_FINISH

$$\frac{P;\Gamma \vdash_p par \triangleright \overline{E} \dashv \Gamma' \qquad P;\Gamma \vdash_p s \triangleright \overline{E}' \dashv \Gamma''}{P;\Gamma \vdash_p \textbf{finish}\,\{par\}; s \triangleright \overline{E}, \overline{E}' \dashv \Gamma''}$$

5.2 Typing of Expressions

Expressions work similarly to the statements above with respect to focusing. We again separate internal/external/focused use for clarity. Looking up a field on the current this uses the internal version of field look-up (fLookupInt) which does not need to widen refined types to owners visible in the interface (E-LOOKUP-FIELD-INTERNAL). Reading a unique local variable requires a destructive read (E-CONSUME-LOCAL), as is the case for accesssing fields and variables during a focus (e.g., CONSUME-FOCUSED-LOCAL). Further, the fields and variables accessed during a focus must be in the currently focused tree.

Expression type rules have the form $P;\Gamma \vdash_p e : T \triangleright E$. Expressions have a type T, and never extend the current Γ. Turnstiles are subscripted by an owner p, which denotes the tree currently being permuted, (i.e., "the tree of p", see Sect. 4.3), lifted to include ϵ (no tree is currently permuted). This subscript controls the reading of variables and fields, and method calls and effects. If we are focusing on the tree of p, then we know that no other part of the system is able to observe these modifications concurrently, and therefore no effects on tree manipulations need to be reported. To preserve tree integrity, all values of the tree must be treated as unique during the focus. The empty effect is denoted ϵ.

$$\boxed{P; \Gamma \vdash_p e : T \rhd E}$$ *(expressions)*

E_LOOKUP_FIELD_INTERNAL
$$\frac{P \vdash \Gamma \quad \Gamma(\text{this}) = B \quad \text{fLookupInt}\,(P, B, f) = T}{P; \Gamma \vdash_\epsilon \text{this}.f : T \rhd \mathbf{rd}\,\mathbf{Rep}}$$

E_CONSUME_LOCAL
$$\frac{p \notin \overline{k} \quad P \vdash \Gamma \quad \Gamma(x) = \mathbf{unique}\,C[\overline{k}]}{P; \Gamma \vdash_p \mathbf{consume}\,x : \mathbf{unique}\,C[\overline{k}] \rhd \epsilon}$$

E_CONSUME_FOCUSED_LOCAL
$$\frac{p \in \overline{k} \quad P \vdash \Gamma \quad \Gamma(x) = C[\overline{k}]}{P; \Gamma \vdash_p \mathbf{consume}\,x : \mathbf{unique}\,C[\overline{k}] \rhd \epsilon}$$

E_NULL
$$\frac{P; \Gamma \vdash T}{P; \Gamma \vdash_p \mathbf{null} : T \rhd \epsilon}$$

E_CONSUME_LOCAL
$$\frac{p \notin \overline{k} \quad P \vdash \Gamma \quad \Gamma(x) = \mathbf{unique}\,C[\overline{k}]}{P; \Gamma \vdash_p \mathbf{consume}\,x : \mathbf{unique}\,C[\overline{k}] \rhd \epsilon}$$

E_CONSUME_FOCUSED_LOCAL
$$\frac{p \in \overline{k} \quad P \vdash \Gamma \quad \Gamma(x) = C[\overline{k}]}{P; \Gamma \vdash_p \mathbf{consume}\,x : \mathbf{unique}\,C[\overline{k}] \rhd \epsilon}$$

E_LOOKUP_FIELD
$$\frac{P \vdash \Gamma \quad \Gamma(x) = T \quad \mathbf{owner}\,(T) = k \quad \text{fLookupRd}\,(P, T, f) = B}{P; \Gamma \vdash_\epsilon x.f : B \rhd \mathbf{rd}\,k}$$

E_LOOKUP_FIELD_INTERNAL
$$\frac{P \vdash \Gamma \quad \Gamma(\text{this}) = B \quad \text{fLookupInt}\,(P, B, f) = T}{P; \Gamma \vdash_\epsilon \text{this}.f : T \rhd \mathbf{rd}\,\mathbf{Rep}}$$

E_CONSUME_FIELD_INTERNAL
$$\frac{P \vdash \Gamma \quad \Gamma(\text{this}) = T \quad p \notin \overline{k} \quad \text{fLookupInt}\,(P, T, f) = \mathbf{unique}\,C[\overline{k}]}{P; \Gamma \vdash_p \mathbf{consume}\,\text{this}.f : \mathbf{unique}\,C[\overline{k}] \rhd \mathbf{wr}\,\mathbf{Rep}}$$

E_CONSUME_FOCUSED_FIELD_INTERNAL
$$\frac{\Gamma(\text{this}) = T \quad \text{fLookupInt}\,(P, T, f) = \mathbf{unique}\,C[\overline{k}] \quad P \vdash \Gamma \quad p \in \overline{k}}{P; \Gamma \vdash_p \mathbf{consume}\,\text{this}.f : \mathbf{unique}\,C[\overline{k}] \rhd \mathbf{wr}\,\mathbf{Rep}}$$

E_UPCAST
$$\frac{P; \Gamma \vdash_{\_} a : T' \rhd \overline{E} \quad P; \Gamma \vdash T' <: T}{P; \Gamma \vdash_p (T)\,a : T \rhd \overline{E}}$$

E_NEW
$$\frac{P; \Gamma \vdash \mathbf{unique}\,C[\overline{k}]}{P; \Gamma \vdash_p \mathbf{new}\,C[\overline{k}] : \mathbf{unique}\,C[\overline{k}] \rhd \epsilon}$$

5.3   Well-Formed Programs

We now present the rules for well-formed declarations. The conditions for well-formed classes in WF-CLASS follows existing ownership systems. Important additions include the helper predicate refinedOwnersUsedOnlyOnce which is a very simple check to see that an owner is neither refined twice, nor does a refined owner appear in more than one field of a class. For example, in the list class where Data = Head ⊎ Tail, the refined owners Head and Tail may only appear *once* on types of fields, including in super classes. This guarantees that permissions are used linearly throughout the fields of the data structure. This is only necessary for fields accessed during a focus, we simplify matters by requiring this always. For simplicity, we assume that names of fields and methods are distinct and that overridden methods preserve types and effects. The appropriate checks are straightforward.

The rules for well-formed fields and methods are standard. We choose to include WF-METHOD and WF-METHOD-FOCUSED to highlight the different requirements of focusing methods. The type environment of a method is that of the class body, plus any owner parameters. The returned variable in the method must have the expected type, and all statements in a method must have a smaller (or equal) effect footprint than what is declared in the method signature.

The focused methods are novel, and are methods that are restricted to operating on a single tree of the data structure. During a focus we may only touch owners refined from the owner p specified in the focuses declaration on the method head. During a focus, the only pointers into p that are accessible are the ones in the current tree, which is why we can omit effects on focusing methods—they work exclusively on data that is not reachable from the rest of the system for the duration of the initial focus block. Inside a focused method, we must cater to the fact that the current `this` may be part of the tree in focus, and also typecheck the entire method body in the current focus, denoted by the subscript p on the turnstiles.

$$\boxed{P \vdash C \,|\, F \,|\, M \quad | \quad P \vdash \textbf{class Object} \dots} \qquad \textit{(Well-formed declarations)}$$

WF_PROGRAM
$$\frac{\forall C \in P.P \vdash C}{\vdash P}$$

WF_OBJECT
$$P \vdash \textbf{class Object} \,[\,\textbf{Owner}\,]\{\}$$

WF_CLASS
$$\frac{\begin{array}{c} \text{distinct}\,(\overline{p}) \\ \forall p \in \overline{p}'.p \in \overline{p} \quad \Gamma = \epsilon, \textbf{Owner} \prec^* \textbf{World}, \textbf{Rep} \prec^* \textbf{Owner}, \overline{p} \succ^* \textbf{Owner}, \overline{K} \\ P; \Gamma \vdash \text{D}[\textbf{Owner}, \overline{p}'] \qquad \text{distinctNames}\,(\overline{F}) \\ \text{refinedOwnersUsedOnlyOnce}(P, \text{C}) \\ P; \Gamma \vdash \overline{F} \qquad \text{distinctNames}\,(\overline{M}) \qquad P; \Gamma, \text{this} : \text{C}[\textbf{Owner}, \overline{p}] \vdash \overline{M} \end{array}}{P \vdash \textbf{class}\, \text{C}[\,\textbf{Owner}, \overline{p}]\, \textbf{where}\, \overline{K}\, \textbf{extends}\, \text{D}[\overline{p}']\{\overline{F}\, \overline{M}\}}$$

WF_FIELD
$$\frac{P; \Gamma \vdash T}{P; \Gamma \vdash T f}$$

WF_METHOD
$$\frac{\Gamma' = \Gamma, \overline{p} \prec^* \textbf{World}, \overline{x : T}^i \qquad P; \Gamma' \vdash \overline{E} \\ P; \Gamma'\, \vdash_{\_}\, s \rhd \overline{E}' \dashv \Gamma'' \\ P; \Gamma'' \vdash_{\_}\, a : T \rhd E \qquad \overline{E}', E \subseteq \overline{E}}{P; \Gamma \vdash [\overline{p}]\, T\, m(\overline{T\, x}^{\,i})\overline{E}\{s;\, \textbf{return}\, a\}}$$

WF_METHOD_FOCUSED
$$\frac{p \in \overline{p} \qquad \text{C}[\textbf{Owner}, \overline{p}]\{\textbf{focused}\, p/p\} = T' \\ \Gamma' = \Gamma, \overline{p}' \prec^* \textbf{World}, \text{this} : T', \overline{x : T}^{\,i} \\ P; \Gamma' \vdash_p s \rhd \overline{E} \dashv \Gamma'' \qquad P; \Gamma'' \vdash_p a : T \rhd \epsilon}{P; \Gamma, \text{this} : \text{C}[\textbf{Owner}, \overline{p}] \vdash [\overline{p}']\, T\, m(\overline{T\, x}^{\,i})\, \textbf{focuses}\, p\{s;\, \textbf{return}\, a\}}$$

5.4 Well-Formed Effects and Effect Disjointness

The rules for well-formed effects and disjointness are found below. Any effect to a valid (possibly refined) owner is a valid effect. For brevity, we omit effect subsumption, which follows existing work in the literature [14], with a straightforward extension for refinements: an effect on p is subsumed by an effect (of the same kind) on q if $q = q_1 \uplus \ldots \uplus q_n$ and $p \in \{q_1, \ldots, q_n\}$. The rules of effect disjointness are standard and follows [14]. Two read effects are always disjoint as read-read races are innocuous.

$$\boxed{P; \Gamma \vdash E \quad | \quad P; \Gamma \vdash \overline{E} \quad | \quad P; \Gamma \vdash E \# E'} \qquad \textit{(Effects and disjointness)}$$

WF_EFF_READS
$$\frac{P; \Gamma \vdash k}{P; \Gamma \vdash \mathbf{rd}\, k}$$

WF_EFF_WRITES
$$\frac{P; \Gamma \vdash k}{P; \Gamma \vdash \mathbf{wr}\, k}$$

WF_EFF_NOTHING
$$\frac{P \vdash \Gamma}{P; \Gamma \vdash \epsilon}$$

WF_EFFS
$$\frac{P; \Gamma \vdash E \qquad P; \Gamma \vdash \overline{E}}{P; \Gamma \vdash E, \overline{E}}$$

FX_MULTI_DISJOINT_MULTI
$$\frac{P; \Gamma \vdash E \,\#\, \overline{E}' \qquad P; \Gamma \vdash \overline{E} \,\#\, \overline{E}'}{P; \Gamma \vdash E, \overline{E} \,\#\, \overline{E}'}$$

FX_DISJOINT_SYMM
$$\frac{P; \Gamma \vdash E_2 \,\#\, E_1}{P; \Gamma \vdash E_1 \,\#\, E_2}$$

FX_SINGLE_DISJOINT_MULTI
$$\frac{P; \Gamma \vdash E \,\#\, E' \qquad P; \Gamma \vdash E \,\#\, \overline{E}}{P; \Gamma \vdash E \,\#\, E', \overline{E}}$$

FX_DISJOINT_NOTHING
$$\frac{P; \Gamma \vdash E_1}{P; \Gamma \vdash E_1 \,\#\, \epsilon}$$

5.5 Well-Formed Types and Subtyping

The rules for well-formed types in WF-TYPE-BASIC are standard ownership rules with the addition of the constraints from the where-clause, which also mention splits and disjointness. Notably, a split on the form $p = p_1 \uplus p_2$ is not a constraint, but rather introduces the disjoint owners p_1 and p_2 in the class body. All well-formed non-unique types have a well-formed unique form (WF-TYPE-UNIQUE).

We allow widening from a refined owner to the owner it refines (SUB-REFINED). This allows taking any link of a linked list and returning it to the list head with the unrefined type List[Rep,D], which allows us to create the last link in Fig. 3 and the rightmost of Fig. 7. Similar to how generics work in Java, such a widening is only sound when reading values, not writing them. (Allowing widening when storing allows the creation of aliases across "disjoint" owners.)

The rules SUB-UNIQUE-FOCUSED and SUB-UNIQUE-REFINED allows us to add a (globally) unique reference to a tree (of locally unique references), both during a focus and not.

$$\boxed{P; \Gamma \vdash T \quad | \quad P; \Gamma \vdash T <: T'}$$ *(Well-formed types and subtyping)*

WF_TYPE_BASIC

$$\textbf{class } C[\textbf{Owner}, \overline{p_i}^i] \textbf{ where } \overline{K} \in P$$
$$\frac{\sigma = \{\textbf{Owner} \mapsto k_1, \overline{p \mapsto k}^i\} \quad \sigma(\overline{K}) = \overline{K}' \quad P; \Gamma \vdash \overline{K}' \quad P; \Gamma \vdash \overline{k}^i \succ^* k_1}{P; \Gamma \vdash C[k_1, \overline{k}^i]}$$

WF_TYPE_UNIQUE

$$\frac{P; \Gamma \vdash B}{P; \Gamma \vdash \textbf{unique } B}$$

SUB_UNIQUE

$$\frac{P; \Gamma \vdash B_1 <: B_2}{P; \Gamma \vdash \textbf{unique } B_1 <: \textbf{unique } B_2}$$

SUB_REFLEXIVE

$$\frac{P; \Gamma \vdash B}{P; \Gamma \vdash B <: B}$$

SUB_DIRECT

$$\textbf{class } C[\textbf{Owner}, \overline{p}^i] \textbf{ where } \overline{K} \textbf{ extends } D[\overline{p}'] \in P$$
$$\frac{P; \Gamma \vdash C[k_1, \overline{k}^i] \quad \sigma = \{\textbf{Owner} \mapsto k_1, \overline{p \mapsto k}^i\} \quad \sigma(D[\overline{p}']) = T}{P; \Gamma \vdash C[k_1, \overline{k}^i] <: T}$$

SUB_TRANSITIVE

$$\frac{P; \Gamma \vdash B_1 <: B_2 \quad P; \Gamma \vdash B_2 <: B_3}{P; \Gamma \vdash B_1 <: B_3}$$

SUB_REFINED

$$\frac{P; \Gamma \vdash C[p \textbf{ from } q, \overline{k}]}{P; \Gamma \vdash C[p \textbf{ from } q, \overline{k}] <: C[q, \overline{k}]}$$

SUB_UNIQUE_FOCUSED

$$\frac{P; \Gamma \vdash C[p, \overline{k}]}{P; \Gamma \vdash \textbf{unique } C[p, \overline{k}] <: C[\textbf{focused } p, \overline{k}]}$$

SUB_UNIQUE_REFINED

$$\frac{P; \Gamma \vdash C[p \textbf{ from } q, \overline{k}]}{P; \Gamma \vdash \textbf{unique } C[q, \overline{k}] <: C[p \textbf{ from } q, \overline{k}]}$$

5.6 Type Environment, Owners and Relations

The type environment Γ is similar to normal ownership types, but additionally uses disjointness assertions and splits, which track how owners are split into disjoint refined owners.

This work introduces two new forms of owners in the ownership literature: p **from** q and **focused** p. The former is a simplification used only in the formalism that allows us to immediately see subset relations (the owner p **from** q keeps track of that p is refined from q). It is not used in the surface language of our prototype implementation. The latter is another simplification visible in REL-FOCUSED—it gives us a simple way to introduce a new owner that copies all relations of an existing owner. The rules for nesting and disjointness are standard. The only new rule is REL-DISJOINT-SPLIT, which allows us to infer disjointness from refinement.

$$\boxed{P \vdash \Gamma \quad | \quad P;\Gamma \vdash k \quad | \quad P;\Gamma \vdash k\,R\,k'}$$ *(Well-formed environment)*

WF_ENV_E
$$\frac{}{P \vdash \epsilon}$$

WF_ENV_VAR
$$\frac{P;\Gamma \vdash T \qquad x \notin \mathsf{dom}\,(\Gamma)}{P \vdash \Gamma, x : T}$$

WF_ENV_REL_INSIDE
$$\frac{P;\Gamma \vdash k_2 \qquad k_1 \notin \mathsf{dom}\,(\Gamma)}{P \vdash \Gamma, k_1 \prec^* k_2}$$

WF_ENV_REL_OUTSIDE
$$\frac{P \vdash \Gamma, k_2 \prec^* k_1}{P \vdash \Gamma, k_1 \succ^* k_2}$$

WF_ENV_REL_DISJOINT
$$\frac{\begin{array}{ll}P;\Gamma \vdash k_1 & \neg\,(P;\Gamma \vdash k_1 \prec^* k_2) \\ P;\Gamma \vdash k_2 & \neg\,(P;\Gamma \vdash k_2 \prec^* k_1)\end{array}}{P \vdash \Gamma, k_1 \,\#\, k_2}$$

WF_ENV_SPLIT
$$\frac{P;\Gamma \vdash q \qquad \forall p \in p_1 \mathinner{..} p_n.\, p \notin \mathsf{dom}\,(\Gamma)}{P \vdash \Gamma, q = p_1 \uplus \mathinner{..} \uplus p_n}$$

WF_ENV_REL_INSIDE
$$\frac{P;\Gamma \vdash k_2 \qquad k_1 \notin \mathsf{dom}\,(\Gamma)}{P \vdash \Gamma, k_1 \prec^* k_2}$$

WF_ENV_REL_OUTSIDE
$$\frac{P \vdash \Gamma, k_2 \prec^* k_1}{P \vdash \Gamma, k_1 \succ^* k_2}$$

WF_OWNER_WORLD
$$\frac{P \vdash \Gamma}{P;\Gamma \vdash \textbf{World}}$$

WF_OWNER_NAME
$$\frac{P \vdash \Gamma \qquad k \in \mathsf{dom}\,(\Gamma)}{P;\Gamma \vdash k}$$

WF_OWNER_REFINED
$$\frac{q = q_1 \uplus \mathinner{..} \uplus q_n \in \Gamma \qquad P \vdash \Gamma \qquad p \in q_1 \mathinner{..} q_n}{P;\Gamma \vdash p \,\mathsf{from}\, q}$$

REL_NEST_TRANS
$$\frac{P;\Gamma \vdash k_1 \prec^* k_2 \qquad P;\Gamma \vdash k_2 \prec^* k_3}{P;\Gamma \vdash k_1 \prec^* k_3}$$

REL_NEST_REFL
$$\frac{P \vdash \Gamma}{P;\Gamma \vdash k \prec^* k}$$

REL_NEST_INSIDE
$$\frac{P \vdash \Gamma \qquad k_1 \prec^* k_2 \in \Gamma}{P;\Gamma \vdash k_1 \prec^* k_2}$$

REL_NEST_OUTSIDE
$$\frac{P;\Gamma \vdash k_2 \prec^* k_1}{P;\Gamma \vdash k_1 \succ^* k_2}$$

REL_DISJOINT_SYMM
$$\frac{P;\Gamma \vdash k_2 \,\#\, k_1}{P;\Gamma \vdash k_1 \,\#\, k_2}$$

REL_DISJOINT
$$\frac{P \vdash \Gamma \qquad k_1 \,\#\, k_2 \in \Gamma}{P;\Gamma \vdash k_1 \,\#\, k_2}$$

REL_DISJOINT_TRANS
$$\frac{\begin{array}{l}P;\Gamma \vdash k_1 \prec^* k_2 \\ P;\Gamma \vdash k_2 \,\#\, k_3\end{array}}{P;\Gamma \vdash k_1 \,\#\, k_3}$$

REL_DISJOINT_SPLIT
$$\frac{\begin{array}{ll} P \vdash \Gamma & q = p_1 \uplus \mathinner{..} \uplus p_n \in \Gamma \\ k_1 \in p_1 \mathinner{..} p_n & k_2 \in p_1 \mathinner{..} p_n \\ & k_1 \not\equiv k_2 \end{array}}{P;\Gamma \vdash k_1 \,\#\, k_2}$$

REL_FOCUSED
$$\frac{P;\Gamma \vdash p \, R \, k}{P;\Gamma \vdash \textbf{focused}\, p \, R \, k}$$

5.7 Dynamic Semantics

We formulate the dynamic semantics as a small-step reduction semantics with parallelism as a non-deterministic choice. Fields are reduced step-wise from left to right, which allows for fine-grained interleaving of statements.

Configurations are on the form $P; \mathcal{D}; \mathcal{H}; \mathcal{F}; \langle e \rangle$ or $P; \mathcal{D}; \mathcal{H};\mathcal{F}; \mathcal{S}$, where P is the static program text, $\mathcal{D}$ maps static owner names p, q to run-time equivalents rk, $\mathcal{H}$ is the heap which maps locations to tuples of $(T, \overline{f \mapsto v})$ and $\mathcal{F}$ which is a stack frame mapping stack variables to the values they store. $\langle e \rangle$ is an expression and $\mathcal{S}$ a stack of statements which capture nesting of async/finish statement and the requirement that all asyncs in a finish be fully reduced before the finish statement is finished (borrowed from [23]). Configurations reduce to other configurations on the same form, or, in the case of a null-pointer dereference, the error configuration **NPError**. For simplicity, we assume the existence of a zero-arity main method in a field-less main class, and the initial configuration has a heap with an instance of this class, the body s of the main method, followed by an assignment into the special variable end, which denotes the program's resulting value.

Inspired by Lightweight Java [37], we maintain a single stack frame ($\mathcal{F}$) for the entire program and take care to uniquify variable names for each method call. The helper $\mathsf{mLookup}_\theta$ used in DYN-DISPATCH performs owner substitution and additionally renames local names to fresh symbols.

Our rules follow standard practises in formalisms with ownership types. For example, owners do not affect a program at run-time, but merely controls what code is considered well-formed and may compile.

Notably, the dynamic semantics of focus is extremely lightweight as the focus block is merely a syntactic trick (DYN-FOCUS). Method call (DYN-DISPATCH) shows the dispatch of a fully reduced call (the receiver and all arguments have been reduced to values). The most involved expression is DYN-NEW which creates a new object on the heap. This creates a new run-time owner for the current Rep and as many new run-time owners as there are refined owners. These are crucial to keep track of the local view internal to an object. Hence, different objects may have different local owners for a single object. These will all be derived from a common "source owner", which is its owning object's Rep, which is key to checking that a local type is consistent with the object's "true" global type. The $\mathcal{D}$ binding set is global for the entire program. Thus, static owner names are distinguished by a subscript denoting the object that "gave rise" to them.

Syntax of run-time elements:

$$\mathcal{H}:: = [] \quad | \quad \mathcal{H}, \iota \mapsto (B, [\overline{f \mapsto v}])$$

$$\mathcal{F}:: = [] \quad | \quad \mathcal{F}, x \mapsto v$$

$$\mathcal{D}:: = [] \quad | \quad \mathcal{D}, k_\iota \mapsto rk$$

$$\mathcal{S}:: = \mathcal{S} \triangleright \mathcal{S}' \quad | \quad \mathcal{S} \| \mathcal{S}' \quad | \quad \langle s \rangle \quad | \quad \sqrt{}$$

Where rk is a run-time owner and a value v are either a location ι or null $\bot$)

$$\boxed{P; \mathcal{H}; \mathcal{F}; \langle re \rangle \hookrightarrow \mathcal{H}'; \mathcal{F}'; \langle re' \rangle} \qquad \qquad \textit{(expressions)}$$

DYN_FIELD

$$\frac{\mathcal{H}(\iota) = (B, \textit{Fields}) \qquad \textit{Fields}(f) = v}{P; \mathcal{D}; \mathcal{H}; \mathcal{F}; \langle \iota.f \rangle \hookrightarrow \mathcal{D}; \mathcal{H}; \mathcal{F}; \langle v \rangle}$$

DYN_FIELD_TARGET

$$\frac{\mathcal{F}(x) = v}{P; \mathcal{D}; \mathcal{H}; \mathcal{F}; \langle x.f \rangle \hookrightarrow \mathcal{D}; \mathcal{H}; \mathcal{F}; \langle v.f \rangle}$$

DYN_LOCAL

$$\frac{\mathcal{F}(x) = v}{P; \mathcal{D}; \mathcal{H}; \mathcal{F}; \langle x \rangle \hookrightarrow \mathcal{D}; \mathcal{H}; \mathcal{F}; \langle v \rangle}$$

DYN_CONSUME_LOCAL

$$\frac{\mathcal{F}[x \mapsto \bot] = \mathcal{F}' \qquad \mathcal{F}(x) = v}{P; \mathcal{D}; \mathcal{H}; \mathcal{F}; \langle \textbf{consume } x \rangle \hookrightarrow \mathcal{D}; \mathcal{H}; \mathcal{F}'; \langle v \rangle}$$

DYN_CONSUME_FIELD_TARGET
$$\frac{\mathcal{F}(x) = v}{P; \mathcal{D}; \mathcal{H}; \mathcal{F}; \langle \textbf{consume } x.f \rangle \hookrightarrow \mathcal{D}; \mathcal{H}; \mathcal{F}; \langle \textbf{consume } v.f \rangle}$$

DYN_CONSUME_FIELD
$$\frac{\mathcal{H}(\iota) = (B, \textit{Fields}) \\ \textit{Fields}(f) = v \qquad \mathcal{H}[\iota.f := \bot] = \mathcal{H}'}{P; \mathcal{D}; \mathcal{H}; \mathcal{F}; \langle \textbf{consume } \iota.f \rangle \hookrightarrow \mathcal{D}; \mathcal{H}'; \mathcal{F}; \langle v \rangle}$$

DYN_NULL
$$P; \mathcal{D}; \mathcal{H}; \mathcal{F}; \langle \textbf{null} \rangle \hookrightarrow \mathcal{D}; \mathcal{H}; \mathcal{F}; \langle \bot \rangle$$

DYN_CAST_SOURCE
$$\frac{P; \mathcal{D}; \mathcal{H}; \mathcal{F}; \langle a \rangle \hookrightarrow \mathcal{D}; \mathcal{H}'; \mathcal{F}'; \langle a' \rangle}{P; \mathcal{D}; \mathcal{H}; \mathcal{F}; \langle (T)\, a \rangle \hookrightarrow \mathcal{D}; \mathcal{H}'; \mathcal{F}'; \langle (T)\, a' \rangle}$$

DYN_CAST
$$P; \mathcal{D}; \mathcal{H}; \mathcal{F}; \langle (T)\, v \rangle \hookrightarrow \mathcal{D}; \mathcal{H}; \mathcal{F}; \langle v \rangle$$

$$\boxed{P; \mathcal{H}; \mathcal{F}; \mathcal{S} \hookrightarrow \mathcal{H}'; \mathcal{F}'; \mathcal{S}'} \hfill \textit{(statements)}$$

DYN_VAR_DECL_INIT
$$\frac{P; \mathcal{D}; \mathcal{H}; \mathcal{F}; \langle e \rangle \hookrightarrow \mathcal{D}; \mathcal{H}'; \mathcal{F}'; \langle e' \rangle}{P; \mathcal{D}; \mathcal{H}; \mathcal{F}; \langle T\, x = e \rangle \hookrightarrow \mathcal{D}; \mathcal{H}'; \mathcal{F}'; \langle T\, x = e' \rangle}$$

DYN_VAR_DECL
$$\frac{\mathcal{F}[x \mapsto v] = \mathcal{F}'}{P; \mathcal{D}; \mathcal{H}; \mathcal{F}; \langle T\, x = v \rangle \hookrightarrow \mathcal{D}; \mathcal{H}; \mathcal{F}'; \langle \textbf{skip} \rangle}$$

DYN_FINISH_ASYNCS
$$\frac{par = \textbf{async } \{s_1\} .. \textbf{async } \{s_n\} \\ \mathcal{S} = \langle s_1 \rangle || .. || \langle s_n \rangle}{P; \mathcal{D}; \mathcal{H}; \mathcal{F}; \langle \textbf{finish } \{par\}; s' \rangle \hookrightarrow \mathcal{D}; \mathcal{H}; \mathcal{F}; \mathcal{S} \rhd \langle s' \rangle}$$

DYN_ASSIGN_FIELD_SINK
$$\frac{\mathcal{F}(x) = v}{P; \mathcal{D}; \mathcal{H}; \mathcal{F}; \langle x.f = e \rangle \hookrightarrow \mathcal{D}; \mathcal{H}; \mathcal{F}; \langle v.f = e \rangle}$$

DYN_ASSIGN_FIELD_SOURCE
$$\frac{P; \mathcal{D}; \mathcal{H}; \mathcal{F}; \langle e \rangle \hookrightarrow \mathcal{D}; \mathcal{H}'; \mathcal{F}'; \langle e' \rangle}{P; \mathcal{D}; \mathcal{H}; \mathcal{F}; \langle \iota.f = e \rangle \hookrightarrow \mathcal{D}; \mathcal{H}'; \mathcal{F}'; \langle \iota.f = e' \rangle}$$

DYN_ASSIGN_FIELD
$$\frac{\mathcal{H}[\iota.f := v] = \mathcal{H}'}{P; \mathcal{D}; \mathcal{H}; \mathcal{F}; \langle \iota.f = v \rangle \hookrightarrow \mathcal{D}; \mathcal{H}'; \mathcal{F}; \langle \textbf{skip} \rangle}$$

DYN_ASSIGN_LOCAL
$$\frac{\mathcal{F}[x \mapsto v] = \mathcal{F}'}{P; \mathcal{D}; \mathcal{H}; \mathcal{F}; \langle x = v \rangle \hookrightarrow \mathcal{D}; \mathcal{H}; \mathcal{F}'; \langle \textbf{skip} \rangle}$$

DYN_CALL_TARGET
$$\frac{P; \mathcal{D}; \mathcal{H}; \mathcal{F}; \langle path \rangle \hookrightarrow \mathcal{D}; \mathcal{H}; \mathcal{F}; \langle path' \rangle}{P; \mathcal{D}; \mathcal{H}; \mathcal{F}; \langle x = path.m[\overline{k}](args) \rangle \hookrightarrow \mathcal{D}; \mathcal{H}; \mathcal{F}; \langle x = path'.m[\overline{k}](args) \rangle}$$

DYN_CALL_ARGUMENT
$$\frac{P; \mathcal{D}; \mathcal{H}; \mathcal{F}; \langle a \rangle \hookrightarrow \mathcal{D}; \mathcal{H}'; \mathcal{F}'; \langle a' \rangle}{P; \mathcal{D}; \mathcal{H}; \mathcal{F}; \langle x = v.m[\overline{k}](\overline{v}\,a\,\overline{a}) \rangle \hookrightarrow \mathcal{D}; \mathcal{H}'; \mathcal{F}'; \langle x = v.m[\overline{k}](\overline{v}\,a'\,\overline{a}) \rangle}$$

DYN_ASSIGN_LOCAL_SOURCE
$$\frac{P; \mathcal{D}; \mathcal{H}; \mathcal{F}; \langle e \rangle \hookrightarrow \mathcal{D}; \mathcal{H}'; \mathcal{F}'; \langle e' \rangle}{P; \mathcal{D}; \mathcal{H}; \mathcal{F}; \langle x = e \rangle \hookrightarrow \mathcal{D}; \mathcal{H}'; \mathcal{F}'; \langle x = e' \rangle}$$

DYN_FOCUS
$$\frac{s'' = (s; s')}{P; \mathcal{D}; \mathcal{H}; \mathcal{F}; \langle \textbf{focus } p \textbf{ in } x \textbf{ as } y \; \{s\} \; ; \; s' \rangle \hookrightarrow \mathcal{D}; \mathcal{H}; \mathcal{F}; \langle s'' \rangle}$$

DYN_SCHEDULING_LEFT
$$\frac{P; \mathcal{D}; \mathcal{H}; \mathcal{F}; S \hookrightarrow \mathcal{D}; \mathcal{H}'; \mathcal{F}'; S'}{P; \mathcal{D}; \mathcal{H}; \mathcal{F}; S \parallel S'' \hookrightarrow \mathcal{D}; \mathcal{H}'; \mathcal{F}'; S' \parallel S''}$$

DYN_SCHEDULING_RIGHT
$$\frac{P; \mathcal{D}; \mathcal{H}; \mathcal{F}; S' \hookrightarrow \mathcal{D}; \mathcal{H}'; \mathcal{F}'; S''}{P; \mathcal{D}; \mathcal{H}; \mathcal{F}; S \parallel S' \hookrightarrow \mathcal{D}; \mathcal{H}'; \mathcal{F}'; S \parallel S''}$$

DYN_ASYNCS_ACTIVE
$$\frac{P; \mathcal{D}; \mathcal{H}; \mathcal{F}; S \hookrightarrow \mathcal{D}; \mathcal{H}'; \mathcal{F}'; S'}{P; \mathcal{D}; \mathcal{H}; \mathcal{F}; S \triangleright S'' \hookrightarrow \mathcal{D}; \mathcal{H}'; \mathcal{F}'; S' \triangleright S''}$$

DYN_ASYNCS_DONE
$$P; \mathcal{D}; \mathcal{H}; \mathcal{F}; \sqrt{} \triangleright S \hookrightarrow \mathcal{D}; \mathcal{H}; \mathcal{F}; S$$

DYN_ASYNC_DONE_LEFT
$$\frac{}{P; \mathcal{D}; \mathcal{H}; \mathcal{F}; \sqrt{} \parallel S \hookrightarrow \mathcal{D}; \mathcal{H}; \mathcal{F}; S}$$

DYN_ASYNC_DONE_RIGHT
$$\frac{}{P; \mathcal{D}; \mathcal{H}; \mathcal{F}; S \parallel \sqrt{} \hookrightarrow \mathcal{D}; \mathcal{H}; \mathcal{F}; S}$$

$$\boxed{P; \mathcal{H}; \mathcal{F}; S \hookrightarrow \textbf{NPError}}$$ (errors)

DYN_NULL_DEREFERENCE_FIELD
$$\frac{}{P; \mathcal{D}; \mathcal{H}; \mathcal{F}; \langle \bot.f \rangle \hookrightarrow \textbf{NPError}}$$

DYN_NULL_DEREFERENCE_UPDATE
$$\frac{}{P; \mathcal{D}; \mathcal{H}; \mathcal{F}; \langle \bot.f = v \rangle \hookrightarrow \textbf{NPError}}$$

DYN_NULL_DEREFERENCE_DISPATCH
$$\frac{}{P; \mathcal{D}; \mathcal{H}; \mathcal{F}; \langle x = \bot.m[\overline{k}](\overline{v}) \rangle \hookrightarrow \textbf{NPError}}$$

5.8   Meta-theory

We prove subject reduction in the normal way, by proving progress and preservation. These theorems contain no surprises, and no unexpected complications appear when sketching their proof.

Progress. For progress, we assert that we can always take one step from a well-formed configuration.

In a well formed P, if $P; \gamma \vdash \mathcal{D}; \mathcal{H}; \mathcal{F}; \mathcal{S}$ then either
$$P; \mathcal{D}; \mathcal{H}; \mathcal{F}; \mathcal{S} \hookrightarrow \mathcal{D}'; \mathcal{H}'; \mathcal{F}'; \mathcal{S}'$$
or
$$P; \mathcal{D}; \mathcal{H}; \mathcal{F}; \mathcal{S} \hookrightarrow \mathbf{NPError}.$$
Similarly for configurations $P; \mathcal{D}; \mathcal{H}; \mathcal{F}; \langle e \rangle$

Preservation. For preservation, we assert that if we can take one step from a well-formed configuration to another, this configuration will also be well-formed.

In a well formed P, if $P; \gamma \vdash \mathcal{D}; \mathcal{H}; \mathcal{F}; \mathcal{S}$ and
$P; \mathcal{D}; \mathcal{H}; \mathcal{F}; \mathcal{S} \hookrightarrow \mathcal{D}'; \mathcal{H}'; \mathcal{F}'; \mathcal{S}'$ then there exists some $\gamma' \supseteq \gamma$ such that $P; \gamma' \vdash \mathcal{D}'; \mathcal{H}'; \mathcal{F}'; \mathcal{S}'$.
Similarly for configurations $P; \mathcal{D}; \mathcal{H}; \mathcal{F}; \langle e \rangle$

Disjointness Invariant. The disjointness invariant asserts that no references in a well-formed heap and stack will break the disjointness assertions of the run-time environment. If a class for example refines an owner X = Y + Z, the run-time representation of Y and Z will always be disjoint as expected.

In a well formed P, if $P; \gamma \vdash \mathcal{D}$ and $P; \gamma; \mathcal{D} \vdash \mathcal{H} \dashv \mathcal{O}$ and $P; \gamma; \mathcal{D} \vdash \mathcal{F} \dashv \mathcal{O}'$ then
$\forall (\iota_1 \in rk_1), (\iota_2 \in rk_2) \in (\mathcal{O}, \mathcal{O}')$.
$\qquad P; \gamma \vdash rk_1 \# rk_2 \implies \iota_1 \neq \iota_2$

Deterministic Parallelism. Finally, the theorem for deterministic parallelism asserts that when starting in a well-formed configuration, if there are more than one way to reach a certain program state, the resulting configurations will be equivalent, *i.e.*, equal up to order and renaming of locations.

In a well formed P, if
$$P; \mathcal{D}; \mathcal{H}; \mathcal{F}; \mathcal{S} \hookrightarrow^* \mathcal{D}'; \mathcal{H}'; \mathcal{F}'; \mathcal{S}'$$
and
$P; \mathcal{D}; \mathcal{H}; \mathcal{F}; \mathcal{S} \hookrightarrow^* \mathcal{D}''; \mathcal{H}''; \mathcal{F}''; \mathcal{S}'$ then $\mathcal{D}' \equiv \mathcal{D}''$, $\mathcal{H}' \equiv \mathcal{H}''$ and $\mathcal{F}' \equiv \mathcal{F}''$

6   Implementation

We have implemented ownership refinement as part of our prototype compiler for our Joelle programming language [12]. Excluding the parser and generated code, the Joelle compiler is about 11 KLOC of Java code using the Polyglot framework [30] and emits Java code. Extending this compiler provides some anecdotal evidence that languages with ownership types can easily be extended with ownership refinement. To add refinement to our compiler, another 1 KLOC of code was necessary.

Joelle is an active-objects language with full ownership-based isolation and safe sharing. Using a well chosen set of defaults and a flat ownership structure the overhead in terms of annotations is very small. The original Joelle language provides coarse-grained parallelism, which can be great for parallelising an existing sequential program by dividing it into a small number of parallel parts wrapped in active objects. This is still one of the main features of Joelle, but we also wanted a language that supports finer-grained parallelism primarily

```
1    owner:Link<D> unlink(int i) focuses D as Q {
2      if (i==1) return dread(next);
3      else return next.unlink(i-1);
4    }
5
6    void sort() writes(rep) {
7      rep:Link<D> tmp = first;
8      focus tmp as rep:Link<Q> {
9        tmp = mergesort(dread(tmp), this.length);
10     }
11     first = tmp;
12   }
```

Fig. 9. unlink and sort methods from Fig. 5 in the current implementation.

for writing parallel programs for multicores from scratch. To this end we have been working on ways to support deterministic parallelism inside active objects. Ownership refinement is one feature designed with this goal in mind, giving "on-demand" rigidity to support deterministic parallelism when required while at the same time providing much of the flexibility programmers are accustomed to otherwise.

The implementation differs slightly from the formalism in superficial ways for practical reasons and because getting things running is more important than nice syntactic sugar and complicated analyses at this stage. As an example, Fig. 9 shows what the unlink and sort methods in Fig. 5 look like in the current implementation. Note the renaming of the focused owner parameter which takes care of any confusion as to which tree we're focusing on in the case where multiple owner parameters are instantiated with the same owner.

7   Related Work

Section 2.3 discussed the shortcomings of introducing additional levels of nesting, adding static regions or unique pointers to express internal structure and use of external values. Figure 1 summarised these problems in a graphical way.

Existing ownership systems such as [1,18,35,38] do not provide a means of introducing structure inside a region, except by using static regions, which are basically just multiple representation owners. This allows a static partitioning of a region, but this is not flexible enough to encode any structure whose size is determined dynamically, and the region partitioning is only accessible inside a single object. Furthermore, regions are global properties, and not local as our refined owners. For example, if elements of a list were partitioned over a number of regions, the same elements could not be shared by some other structure.

In Clarke and Drossopoulou's ownership-based effect system [14] and derivatives, the smallest effect is to an entire region, which makes perfect sense as this system is unable to distinguish *e.g.*, two links belonging to the same list. Ownership refinement naturally introduces increased precision up to reads or writes to one specific object (in the form of a permission set of size 1) or subregion.

This is key to implementing structures with internal parallelism, regardless of whether the objects involved are internal or external to the data structure, or both.

Viewing owner parameters as sets of reference permissions was explored in the first author's Master Thesis [8].

Parallel Programming. Ownership systems have been applied to parallel programming or similar problems (*e.g.,* [16,25,27]. Extant proposals are unable to express the example in Fig. 3 as they cannot statically capture mutual alias-freedom for all list elements. Introducing uniqueness (*e.g.,* in the style of Boyland et al. [6,7]) and making the element pointer unique has the side-effect of preventing all non-local aliases of the elements, but is also not enough on its own since traversal of the **next**-links must be acyclic. Making the **next** fields unique also would solve this problem, but would prevent several useful scenarios including the last link or doubly-linked list in Fig. 7, which our system supports either by using a separate tree for the last and previous pointers, or by using non-refined references for these pointers, since one tree is probably enough in this particular example.

Refined ownership has more uses than static checking of deterministic parallelism. Nevertheless, several such works, most notably Deterministic Parallel Java, have invented related machinery. Therefore, it is relevant to discuss several such works here. The following ownership systems are designed with parallelism in mind, but suffer from the problem overcome by this paper: it is not possible to statically distinguish between references owned by the same object.

Cunningham *et al.* [17] constructed a system based on universe types [29] for race safety where disjointness is only introduced by ownership nesting. Craik and Kelly [16] employ ownership types to infer parallelism in C# programs. To overcome the coarse grained nature of ownership, they additionally perform dynamic alias checking (*e.g.,* comparing all pointers in a collection to make sure they do not alias) or rely on programmer assertions.

Lu *et al.* [25,26] use effects on ownership contexts to infer synchronisation requirements for parallel tasks and perform lock correlation. Where they suffer from the coarse-grainedness of ownership (*e.g.,* all external pointers in a list, all links internal to the list) they infer or must use locks to guarantee that no critical section has more than one active thread. As a consequence, a parallel map over a list requires locking on each individual element to avoid potential races due to aliasing. Lu and Potter define a type system for acyclicity [24] that partitions the heap into regions and constrain references between regions to go in one direction. They can eliminate cycles between a link and its respective element, but fall short to eliminate cycles in a chain of links, or duplicates in a list.

Milanova and Huang [28] develop a static analysis tool for detecting an over-approximation of data races based on dominance in the object graph. No fine-grained partitioning is supported.

The work on Deterministic Parallel Java [2,4] offers a rich set of tools for deterministic parallel programming, and its region path lists are very similar to ownership regions. Deterministic Parallel Java supports both vertical nesting (A owns B) and horizontal nesting (A is divided into a static number of regions). In [3] DPJ is extended to support external data elements in a way similar to ours, however the correctness of parallel traversals is not guaranteed by the system but left as an exercise to the programmer. The only way to initially store elements in a region with disjointness is by sending object factories into the objects that define them, *i.e.,* let the list populate itself. As a consequence, it is not possible to create a value elsewhere in the system and later move it into *e.g.,* a list. This is rectified in later work [5], where the concept of a reference group is introduced which allows tree-like traversal of data structures as long as traversal stays within the reference group. This is similar to refined ownership where trees are in a constant focus. Special effects are used to preserve tree integrity, and populate empty reference groups from other groups.

Servetto *et al.* [36] extend Ballon Types for safe parallelisation over arbitrary object graphs. This work is strongly related to that of Gordon *et al.* [20]. They can safely encode alias-freedom for the nodes in binary trees, but are unable to do so in the presence of back-pointers (*e.g.,* doubly-linked lists) or multiple entry points into a structure (last pointers), or pointers to unrestricted external data.

In ongoing work (submission 44), Brandauer *et al.* employ a type system that allows to gradually relax uniqueness of references to express similar information as this paper does. Refined Ownership differs in that it takes a fundamentally different starting point and, being a straightforward extension, integrates well with other proposals based on ownership types.

8   Conclusion

Refined ownership extends ownership types with the possibility to distinguish between objects owned by a single owner. An important aspect of refinement is that it is a local concern that does not leak to the outside world: refinements introduce fine-grained reasoning without forcing clients of objects that use them to propagate additional ownership information. The main driver behind this work was to type structured parallelism inside active objects, but the usefulness of ownership refinement extends beyond this use case.

On-going work generalises refined ownership to more set operations on owners-as-sets-of-permissions. This allows *e.g.,* building lists of elements owned by different owners by constructing an owner as the their union, etc. This increases the flexibility of ownership types in a way that avoids flattening the ownership hierarchy or introduce gratuitous copying of objects across owner boundaries.

Refined ownership can express internal structure, alias-freedom and non-interference inside data structures in ways that normal ownership systems are unable to do, and importantly allow local reasoning about references to external values. In a system that already uses ownership types, their addition is very small, only imposes syntactic overhead where used, and adds considerable reasoning power.

References

1. Aldrich, J., Chambers, C.: Ownership domains: separating aliasing policy from mechanism. In: Odersky, M. (ed.) ECOOP 2004. LNCS, vol. 3086, pp. 1–25. Springer, Heidelberg (2004)
2. Bocchino, R.: An effect system and language for deterministic-by-default parallel programming, 2010. Ph.D. thesis, University of Illinois at Urbana-Champaign (2010)
3. Bocchino Jr, R.L., Adve, V.S.: Types, regions, and effects for safe programming with object-oriented parallel frameworks. In: Mezini, M. (ed.) ECOOP 2011. LNCS, vol. 6813, pp. 306–332. Springer, Heidelberg (2011)
4. Bocchino, R., Adve, V.S., Dig, D., Adve, S.V., Heumann, S., Komuravelli, R., Overbey, J., Simmons, P., Sung, H., Vakilian, M.: A type and effect system for deterministic parallel Java. In: OOPSLA, pp. 97–116 (2009)
5. Bocchino, R., Aldrich, J.: Reference groups for local uniqueness. Technical report, CMU (to appear)
6. Boyland, J.: Alias burying: unique variables without destructive reads. Softw. Pract. Exp. **31**(6), 533–553 (2001)
7. Boyland, J.T., Retert, W.: Connecting effects and uniqueness with adoption. In: POPL, pp. 283–295 (2005)
8. Castegren. E.: Laps : a general framework for modeling alias management using access permission sets, Master thesis (2012)
9. Cavé, V., Zhao, J., Shirako, J., Sarkar, V.: Habanero-java: the new adventures of old x10. In: Proceedings of the 9th International Conference on Principles and Practice of Programming in Java, PPPJ 2011, pp. 51–61. ACM, New York (2011)
10. Clarke, D.: Object ownership and containment. Ph.D. thesis, School of Computer Science and Engineering, University of New South Wales, Australia (2002)
11. Clarke, D., Wrigstad, T.: External uniqueness is unique enough. In: Cardelli, L. (ed.) ECOOP 2003. LNCS, vol. 2743, pp. 176–200. Springer, Heidelberg (2003)
12. Clarke, D., Wrigstad, T., Östlund, J., Johnsen, E.B.: Minimal ownership for active objects. In: Ramalingam, G. (ed.) APLAS 2008. LNCS, vol. 5356, pp. 139–154. Springer, Heidelberg (2008)
13. Clarke, D., Östlund, J., Sergey, I., Wrigstad, T.: Ownership types: a survey. In: Clarke, D., Noble, J., Wrigstad, T. (eds.) Aliasing in Object-Oriented Programming. LNCS, vol. 7850, pp. 15–58. Springer, Heidelberg (2013)
14. Clarke, D.G., Drossopoulou, S.: Ownership, encapsulation and the disjointness of type and effect. In: OOPSLA, pp. 292–310 (2002)
15. Clarke, D.G., Potter, J., Noble, J.: Ownership types for flexible alias protection. In: OOPSLA, pp. 48–64 (1998)
16. Craik, A., Kelly, W.: Using ownership to reason about inherent parallelism in object-oriented programs. In: Gupta, R. (ed.) CC 2010. LNCS, vol. 6011, pp. 145–164. Springer, Heidelberg (2010)
17. Cunningham, D., Drossopoulou, S., Eisenbach, S.: Universes for Race Safety (2007)
18. Dietl, W.M.: Universe Types: Topology, Encapsulation, Genericity, and Tools. Ph.D., Department of Computer Science, ETH Zurich, Doctoral Thesis ETH No. 18522, December 2009
19. Fähndrich, M., DeLine, R.: Adoption and focus: practical linear types for imperative programming. In: PLDI, pp. 13–24 (2002)
20. Gordon, C.S., Parkinson, M.J., Parsons, J., Bromfield, A., Duffy, J.: Uniqueness and reference immutability for safe parallelism. In: OOPSLA, pp. 21–40 (2012)

21. Greenhouse, A., Boyland, J.: An object-oriented effects system. In: Guerraoui, R. (ed.) ECOOP 1999. LNCS, pp. 205–229. Springer, Heidelberg (1999)

22. Igarashi, A., Pierce, B.C., Wadler, P.: Featherweight java: a minimal core calculus for Java and GJ. ACM Trans. Program. Lang. Syst. **23**(3), 396–450 (2001)

23. Lee, J.K., Palsberg, J.: Featherweight x10: a core calculus for async-finish parallelism. In: PPOPP, pp. 25–36 (2010)

24. Lu, Y.: A type system for reachability and acyclicity. In: Gao, X.-X. (ed.) ECOOP 2005. LNCS, vol. 3586, pp. 479–503. Springer, Heidelberg (2005)

25. Lu, Y., Potter, J., Xue, J.: Ownership types for object synchronisation. In: Jhala, R., Igarashi, A. (eds.) APLAS 2012. LNCS, vol. 7705, pp. 18–33. Springer, Heidelberg (2012)

26. Lu, Y., Potter, J., Xue, J.: Structural lock correlation with ownership types. In: Felleisen, M., Gardner, P. (eds.) ESOP 2013. LNCS, vol. 7792, pp. 391–410. Springer, Heidelberg (2013)

27. Lu, Y., Potter, J., Zhang, C., Xue, J.: A type and effect system for determinism in multithreaded programs. In: Seidl, H. (ed.) ESOP 2012. LNCS, vol. 7211, pp. 518–538. Springer, Heidelberg (2012)

28. Milanova, A., Huang, W.: Static object race detection. In: Yang, H. (ed.) APLAS 2011. LNCS, vol. 7078, pp. 255–271. Springer, Heidelberg (2011)

29. Müller, P., Poetzsch-Heffter, A.: Universes: a type system for controlling representation exposure. In.Fernuniversität Hagen Programming Languages and Fundamentals of Programming (1999)

30. Nystrom, N., Clarkson, M.R., Myers, A.C.: Polyglot: an extensible compiler framework for java. In: Hedin, G. (ed.) CC 2003. LNCS, vol. 2622, pp. 138–152. Springer, Heidelberg (2003)

31. Östlund, J., Brandauer, S., Wrigstad, T.: The joelle programming language : evolving java programs along two axes of parallel eval. In: LaME 2012 (2012)

32. Östlund, J., Wrigstad, T.: Regions as owners - a discussion on ownership-based effects in practice. In: IWACO 2011, International Workshop on Aliasing, Confinement and Ownership in Object-Oriented Programming (2011)

33. Östlund, J., Wrigstad, T.: Multiple aggregate entry points for ownership types. In: Noble, J. (ed.) ECOOP 2012. LNCS, vol. 7313, pp. 156–180. Springer, Heidelberg (2012)

34. Östlund, J., Wrigstad, T., Clarke, D., Åkerblom, B.: Ownership, uniqueness, and immutability. In: Paige, R.F., Meyer, B. (eds.) TOOLS. LNCS, pp. 178–197. Springer, Heidelberg (2008)

35. Potanin, A., Noble, J., Clarke, D., Biddle, R.: Generic ownership for generic Java. In OOPSLA, pp. 311–324 (2006)

36. Servetto, M., Pearce, D. J., Groves, L., Potanin, A.: Balloon types for safe parallelisation over arbitrary object graphs. In: 4th Workshop on Determinism and Correctness in Parallel Programming (2013)

37. Strnisa, R., Parkinson, M.J.: Lightweight java. In: Archive of Formal Proofs (2011)

38. Wrigstad, T.: Ownership-Based Alias Management. Ph.D. thesis, Royal Institute of Technology, Kista, Stockholm, May 2006

Author Index

Printed in the United States
By Bookmasters